The End of Fossil Energy

and

The Last Chance
for Survival

McIntire
· PUBLISHING SERVICES ·

The publisher gives permission for any part of this publication to be reproduced with appropriate reference and not for profit and/or resale.

ISBN 0-9743404-3-X

First Edition, Published 2003
Second Edition, Published 2005
Third Edition, Published 2006

The editors for this book were Dr. Don McLean and Debbie Howe. The cover art and interior layout was done by Virginia Howe. It was set in Janson by McIntire Publishing.

For more information please contact the author at Howe Engineering Company, 298 McIntire Road, Waterford, ME 04088 or McIntire Publishing Services through their website at

www.mcintirepublishing.com

The End
of Fossil Energy

and

The Last Chance
for Survival

by John G. Howe

McIntire
· PUBLISHING SERVICES ·

Dedication

To my children and all the children of the world—
they will need all the resources we can conserve.

Contents

Appendices

Acknowledgments

I would like to directly acknowledge the leadership of Dr. Colin Campbell, the founder of the Association for the Study of Peak Oil and Gas (ASPO). You can download a wonderful picture of this distinguished Irish gentleman as well as the following quotation from the ASPO website **www.peakoil.net**:

> *Understanding depletion is simple. Think of an Irish pub. The glass starts full and ends empty. There are only so many more drinks to closing time. It's the same with oil. We must find the bar before we can drink what's in it.*

If the world survives the looming energy crisis, it will largely be because of this one man's efforts.

*   *   *

On a personal level, three people have been indispensable during the preparation of this book:

Don McLean, D.V.M., a lifelong, alternative lifestyle and renewable-energy enthusiast who practices what he preaches. Don has been a hands-on mentor for all things relating to energy as well as number one critic for my English composition.

Virginia Howe, my daughter, a career graphic artist and publisher. Her knowledge allowed me to transform my manuscript into a printed book.

Debbie Howe, my loving wife, patient sounding board, my keyboard link to this book and the outside world.

Preface to the First Edition

I'm sitting in a rural New England farmhouse in February 2003 trying to start this project for the fifth time in the last 10 years. The setting could not be more ideal in my comfortable, senior years as a retired engineer, farmer, part-time historian, and sometimes skeptic. Yet, something seems terribly wrong. It's a cold winter day, and in the last few minutes the back-up oil burner has switched on to smooth out the heat from our wood-stoves. Now, silence is disturbed again by the school bus delivering the neighbor's children. A huge logging truck just trundled down the road with wood destined for who knows where—it varies, pulp for one of the paper mills, saw logs for Japan, or maybe just biomass for chipping and electrical cogeneration. A skidder is still roaring into the late afternoon, and a jet has left the east coast swinging west overhead. Fortunately, the snowplow is only sanding today, and our driveway is clear thanks to a gallon of gas in our 50-year old John Deere tractor.

The sound of energy intrudes, as does the TV reporting the latest on the pending war in Iraq (which just happens to have over 100 billion barrels of conventional oil reserves, the second largest in the world after Saudi Arabia). The war news is frequently interrupted by ads for things we must have, like 4 x 4 pick up trucks and snowmobiles.

Our president mentioned hydrogen fuel in his State of the Union address the other night. It's hard for me, as an engineer, farmer, and manufacturer of bicycle-powered generators, to imagine a hydrogen-powered school bus, skidder, snowplow, tractor, jet plane, etc. I am sure he means well but wonder if he understands the language of BTU's, watts, kilowatt hours, and calories as well as the details of non-fossil energy sources like solar, nuclear, wind, geothermal, biofuels, and hydro.

There is a hard-to-resist pressure to consume and throw away because our fast-paced materialistic lifestyle is built on a foundation of very low-cost fuel. From consumer goods shipped from China to a supermarket full of food or a tank of gasoline, energy is an integral part of our lives. Yet, we know intuitively, and many experts are telling us, this present epoch cannot last much longer. There are many sources of energy to do our work (other than food which fuels our muscles). The fossil fuels are inexpensive and supply 86% of our needs, but they exist in limited quantities. Renewable-energy resources, like solar and wind, will last forever but presently contribute only a tiny part of our total consumption.

All the things I've mentioned are small pieces of a huge puzzle. The picture becomes much clearer after years of collecting and studying the pieces. The emerging composite is so ominous and frightening that I feel compelled to speak out. I must try to reach as many people as possible while there is still time to act and there are remaining resources to facilitate change. This is a compelling story that concerns everyone and needs to be told. There are many similar voices with more credentials and better words than mine, but they are on the web or in obscure, technical, and expensive books. One thing I will try to do differently is to condense the energy story into one small book, which will summarize and bring to date the subject in a manageable form. I will self-publish the first three-thousand copies for free distribution and attempt to reach as many as possible with the facts. Please read on and get involved as an informed citizen demanding answers, action, and who will vote for equally concerned leaders to get us moving in the right direction.

The astonishing communication power of the World Wide Web is unique to our age. Electronic information may be a key source for quickly providing facts, resources, and opportunities for public involvement. If you have any doubts concerning the coming crisis, I suggest you first turn to Chapter 3, which lists an ever-growing number of pertinent websites and a ground swell of concern about the demise of fossil fuels. Unfortunately, only a fraction of the public uses the web. Too many rely on simplistic TV, which does not focus on scary projections, especially those that might offend advertisers or consumer-market segments.

By the spring of 2004, this energy book will be published. The public has to become involved and time is of the essence. On August 14, 2003, a series of obscure "natural" (does this mean non-terrorist?) events shut down the electricity in the eastern one-third of the United States from New York City to Cleveland to Canada. Officials scrambled to find what went wrong and how to prevent it from reoccurring. This event serves to show how important just the electrical component of energy is in our complex society, yet most of the American public is still unaware of the facts showing the 100-year epoch of low-cost energy is clearly entering its twilight period. This will surely lead us to a much different future. More power plants and transmission lines are not the solution without fuel to feed them. Honest energy legislation should first focus on long-term availability of basic fuel sources.

Every citizen of the modern world should know the story of energy because everyone is a participant. The outcome will depend on all of us working together. Cataclysmic times are at our doorstep. We, our children, and our grandchildren cannot expect to continue the energy intoxication of the 20th century.

Post-Preface

The epilogue to the story of fossil energy cannot be written for another fifty years. By then I would be 120 years old so I probably won't be here. At the present time, depletion of oil, natural gas, and coal seem to be a silent topic, at least in the USA. Could this be intentional because the subject is so horrific or just denial? Obscure websites or offhand mention in the media relegates energy to the status of dubious science or conspiracy theories. For example, the Jan 22, 2004 issue of *The New York Times* discusses the January 9th announcement by Shell that they are drastically reducing their oil-proven-reserve estimate by 3.9 billion barrels. Instead of mentioning the possibility of world depletion, the article focuses on investor "significant concern and in some quarters, outrage" against management and the form of corporate structure.

The status quo is extremely discouraging considering the small-time frame we have for the changes we must make. The next several years will show if we are up to the task.

Preface to the Second Edition

The Last Chance for Sustainability

It is now February 2005, two years after starting the first edition of *The End of Fossil Energy and a Plan for Sustainability*. Since then the world has consumed another 55 billion barrels of oil, including 15 billion in the U.S., which is twice the 7 billion expected to be in ANWR. The Market price has doubled to almost 60 dollars per barrel. Oil company profits are at record highs but far fewer oil fields are being discovered each year. There are previously discovered fields coming into production and about one-half of the world's original endowment is still in the ground, but worldwide demand has clearly outpaced production.

China and India have experienced unexpectedly high economic expansion and have become major players in the world energy markets. Due to continued violence, Iraq is only producing about 1.9 million barrels of oil per day, well below its pre-war level of 2.5 million.

General Motors stock has plummeted over 50% to the mid-twenty range, and 460,000 retirees may be depending on a very shaky pension fund. In total, GM's 300 billion dollar long-term debt is massive compared to a market capitalization of about 16 billion and expected 2005 loss of over 2 billion dollars. GM's (and Ford's) debt ratings have dropped to junk bond investment levels. Watch the price of oil and GM's stock price daily to see how our country fares. Clearly these are precarious times.

The subtitle of the second edition for this book has been changed to *The Last Chance for Sustainability*. The third edition (if ever) might be subtitled *The Last Chance for Survival*.

On a positive note, in the last year, many new energy-related websites have come on line and many new books published. America is waking up to peak oil and fossil-fuel depletion. Our personal initial printing of 3,000 books has been sent out all over the world. Hopefully we've alerted at least a tiny number of people. Our national and world energy dilemma has become more focused for more people.

In the interest of cost and time, the first edition is herein reprinted intact except for editorial corrections and clarification. The only major change in the second edition is a new Chapter 2, "A Two-Year Update 2005" and the move of "Energy 101" to Appendix 6.

After a year's positive experience with our solar-powered tractor (SPUV), we have expanded our thoughts about this promising concept into a new Appendix 7.

While reflecting on the whole subject of fossil energy depletion, several undeniable conclusions visibly surface:

1. World fossil energy with consumption led by the U.S. (the Saudi Arabia of the mid-twentieth century) is clearly approaching terminal depletion, possibly faster than expected. Remaining coal (with excessive CO_2 emissions) will be the dominant remaining fuel as we decline from our peak of fossil fuels and concurrent industrialized civilization.

2. There are no similar substitutes for high-energy fossil fuels except for miniscule quantities of biofuels, which also begin with sunlight and can concentrate and store energy in a few years instead of hundreds of millions. The problems of bioenergy for use as fuel instead of food are discussed at length in the new Chapter 2.

3. Geopolitically, the world is shifting uneasily because of the scramble to source, isolate, transport, speculate, and buy and sell fossil fuels for the best short-term motives of individuals, consumers, and producer nations. Oil and gas reserves in the U.S. and North America are seriously in decline. Resource destabilization is obvious in the Middle East, the Far East, Russia and former Soviet Union coun-

tries, Africa . . . everywhere in the world except maybe Iceland, which is uniquely endowed with copious geothermal power. Don't let anything fool you, "It is about oil."

4. If we have any chance to survive the mess we're in, we must start immediately with conservation. I'm not talking about baby steps with hybrids or fluorescent bulbs. We need HUGE changes in direction from our present trend of growth and consumption. The fuel we save now might provide a softer landing and, hopefully, survival for our children. Conservation, led by the U.S., would also relax the pressure and world tension.

5. We need to quickly spread the message of fossil-fuel depletion to the American public and stop assuaging them with confusion and diversions like hydrogen. Our only hope in the short time we have is to somehow reach the general population as well as our elected officials and media and make energy the number one subject.

You, as an individual reader, can help disseminate the message of fossil-energy depletion, but only if you become involved and make it happen. Hopefully our small book will help.

THE LAST CHANCE FOR SUSTAINABILITY

One last time, to summarize our only hope to sustain any semblance of an acceptable Industrialized World, I offer the following Five Percent Plan as covered in detail in this book. I realize there may be some who do not wish to believe in or follow a program to **save civilization**, but remember, in a democracy, we only need a majority to enact legislation that all must follow. We don't need to reach every single person.

1. **Convince yourself** and others that peak oil and fossil fuel depletion are real, and at our doorstep. Denial is contrary to facts, math, and science. Many voices have been trying to warn us for years. Now we are at the tipping point.

2. Argue that the looming energy crisis be recognized as the **single most important challenge** and subject facing us. There are no other issues more pertinent to our and our children's survival.

3. **Support leaders** who recognize our non-renewable energy addiction, will enact appropriate legislation, and educate the public why this is necessary. Remember, we don't need to convince everyone. In a democracy, we only need a majority for the rule of law to control our destiny.

4. Our first immediate move is to enact **energy rationing**. Concurrently, massive-education programs are necessary to convince the public why these measures are absolutely necessary for future survival. Market forces and price alone will not work. Everyone, rich and poor, should share the effort. We understood rationing in World War II. We are now at an even more serious time.

5. Our second move is to begin an immediate transition to the two high-tech **renewable energies** that can be expanded in large scale, specifically solar and wind. Other non-fossil sources are finite (like geothermal, and nuclear) or mirages (like hydrogen and biofuels).

6. Initiate a public relations campaign to not only explain our dire situation but why the only basic and ultimate hope to save civilization in the long run is to **reduce population**. If we avoid this subject, we are truly doomed to die off like any other living species that outstrips resources. (See the new book, *Collapse*, by Jared Diamond and reviewed in Chapter 2.)

7. Our fourth move is to **reduce our military presence**. We will need all of our resources to survive as a nation without being a contestant and policeman for declining world energy.

8. Finally, we need to deglobalize, decentralize, and **revert to community solutions** with neighborhood economies of approximately a 25-mile radius, the distance of travel by a bike or solar-electric vehicle. The community periphery will be devoted to hands-on agriculture to feed primarily

that community. The village core will provide consumer services, energy storage, and specialty skills including battery recycling. The community center will also be the social nucleus and departure point for electric-mass-public transportation to other communities as well as national centers of commerce and government. (See **www.communitysolution.org**).

Preface to the Third Edition

The Last Chance for Survival

Peak Oil

In less than one year since our second printing in 2005, it is apparent that the subject of fossil-energy has moved into the mainstream of public awareness. The following items highlight the transition of energy topics from a fringe group of gloom and doom alarmists to accepted legitimacy.

The term peak oil has become the common phrase for media and public attention. In the 12 months from January 1, 2005 to January 1, 2006 crude oil jumped 36 % from $45.00 per barrel to $61.00. By January 15, 2006 it had increased another 9% to $67.50. In the same period, natural gas increased a whopping 100% from the $6.00 range to over $12.00 per thousand cubic feet. By January 25, 2006 it had settled back to $8.25 due in large part to the extraordinarily mild winter in the U.S.

On December 8, 2005 the House Energy and Commerce Subcommittee on Energy and Air-Quality held the first full-scale congressional hearing on peak oil. This bipartisan caucus is co-chaired by Roscoe Bartlett (R-Maryland) and Tom Udall (D-New Mexico). Resolution 507 was co-sponsored by 16 prominent congressmen and begins with the first paragraph:

> *Expressing the sense of the House of Representatives that the United States, in collaboration with other international allies, should establish an energy project with the magnitude, creativity, and sense of urgency that was incorporated in the "Man on the Moon" project to address the inevitable challenges of Peak Oil.* (See **www.energycommerce.house.gov** for a complete transcript.)

The first ASPO-USA (Association for the Study of Peak Oil and Gas) conference was held November 10–11, 2005 in Denver, Colorado. With a packed audience of 400 and hosted by Denver's mayor, John Hickenlooper, this meeting clearly showed a growing mainstream public concern over the future of energy. The concept of sustainable growth was brought to the forefront by former Colorado Governor Richard Lamm. See **www.aspo-usa.org** for transcripts.

On December 1, 2005 geologist Kenneth Deffeyes ("Hubbert The View from Hubbert) gave a lecture at Cal Tech to a packed audience of 800 people. One of his quotes, "I'm not so worried about the next 15 to 20 years. I'm very scared about the next five," set the tone for his entire talk.

The second largest oil corporation, Chevron, has purchased multi page ads in major publications like *Newsweek, Scientific American, The Economist* calling explicit attention to the coming oil crisis. See **www.willyoujoinus.com.**

The *Financial Sense News Hour* proclaims energy, the big story of 2005. The transcript of this December 24, 2005 comprehensive analysis can be found on **www.financialsense.com**. It is no longer possible to keep up with the rapidly increasing number of conferences, reports, media publications, new books and web sites discussing the enormity and crisis implications of peak oil.

The Triple-Crisis ("Peak Iceburg")

But wait! Peak oil is only the most visible tip of our growing dilemma. The challenge facing civilization becomes more ominous. For years, respected authors and lecturers have been attempting to raise public concern about two other inextricably related issues: population and ecology. Unfortunately, during the last century of fossil energy excess, very few would listen and the situation has grown progressively worse. Consumption and population continued to increase in lock-step (as would be expected with any biological species in an environment of temporarily unlimited energy resources) with the devastation of the earth's natural systems.

Population

In 1798, the classic work, *An Essay on the Principle of Population*, was written (at first anonymously) by. T. R. Malthus. His main thesis is that the natural trend for population to increase exponentially can never be matched by finite resources. The tension between these two forces is destined to end in misery. But, Malthus did not know of the potential of concentrated fossil energy plus concurrent advances in modern medicine and agriculture that temporarily proved him wrong. In the two hundred years since his warning, population increased from a few hundred million to almost 7 billion. Today, most of the world is sliding over the cliff of survival while a minority in the industrialized (intellectual?) world enjoys the end of the energy party ... drunk on oil and blind to the hangover soon to come. (Even Pres. Bush in his Jan. 2006 state of the union speech claimed we were addicted to oil.)

A few voices have been anticipating the overshoot of population for years. Respected authors and academic professors Albert Bartlett, *The Essential Exponential;* Paul Ehrlich, *Population Bomb, The Stork and the Plow;* Kenneth Smail, *Remembering Malthus I, II, III;* Lindsey Grant, *The Collapsing Bubble, Too Many People;* Joel Cohen, *How Many People Can the Earth Support;* Dennis Meadows, *Limits to Growth: The 30 Year Update* have made the case that world population has far exceeded the limits of sustainable resources. The most comprehensive and compelling of all these population books is William Stanton *The Rapid Growth of Human Populations* 1750–2000. Stanton points out unequivocally, with completely researched numbers, how the world and each country tend to increase in population in good times when WROG (Weak Restraint on Growth) is in control until the maximum carrying capacity of over-populated regions is reached. At this point, the VCL (Violent Cutback Level), local collapse and chaos usually occur until, hopefully, a new lower level of population can regain traction and repeat the process all over again.

Until peak oil, the public would have no ear for these Cassandras. The urgency was not there, future generations can worry about their own fate. The critics extrapolated from past trends to invent the oxymoron of future sustainable growth.

The public ear for optimism always trumped the Chicken Littles. The optimists argue that when the time comes, our superior intelligence will invent new technology to keep the party going. Or, they argue that since a very few modern countries led by Russia, Italy, and Japan have achieved negative population growth, it can be done when necessary. Or, since negative growth will have such a deleterious impact on our economic system, it is unthinkable that it will happen.

Eco-Devastation

A third group of authors, educators, and advocacy groups have been beating a long, background drum-roll of ecological concerns. The forefront of these anxieties has been clear evidence of global warming, or, more correctly, climate change. The CO_2 emissions from fossil fuel combustion have, as would be expected, increased in lock-step with increased consumption and population growth. The documented melting of ice caps and glaciers, combined with unprecedented hurricanes spawned over 90°f water temperatures, are rapidly gaining public attention.

There are dozens of other obvious examples of drastic ecological decline. Top-soil loss, forest devastation, clean water availability, species extinction, atmospheric and soil pollution, all have their vociferous champions. All are obviously directly related to fossil fuel consumption and population growth.

The respected voices of eco-devastation and its historic effect on societies are exemplified by: Jarred Diamond in his book, *Collapse;* Thomas Homer-Dixon, *Environment, Scarcity, and Violence;* David and Marcia Pimmental, *Food, Energy, and Society;* Walter Youngquist, *GeoDestinies: The Inevitable Control of Earth Resources Over Nations and Individuals;* Clive Ponting, *A Green History of the World: the Environment and the Collapse of Great Civilizations;* and Lester Brown, *Plan B; Rescuing a Planet Under Stress and a Civilization in Trouble.* Like population, these comprehensive, quantitative messages have been marginalized to the fringes of envirogroups. "We've heard that before, but the party's still going."

The combined inertia of short-term corporate profit with more growth as the necessary support for further endless growth and feel-good political savvy, all conspire to overwhelm the eco-voices.

Enter Peak Oil

The clear awakening of public perception about the reality of the imminent fossil energy crisis casts new meaning for the other two long-simmering areas of concern, population and eco-devastation. Combined, the three interrelated subjects form the critical mass (or perfect storm) and predict the collapse of modern civilization. This will be no different, just quicker than preceding historic collapses where population out-stripped resources leading to the four apocalyptic horses: famine, pestilence, war, and death.

One would think, with our modern technology combined with unique communication skills and clear knowledge of history, that we could be the first biological species to take control of our fate. We spend billions searching for solutions to health problems or looking for and suggesting ways to deflect NEOs (Near Earth Objects) that might impact the earth, yet ignore the greatest threats of all.

Unfortunately, the combined urgency and synergism of the three world crisis components (energy, population, eco-devastation) indicate that we may already have passed the point of survival. The most convincing of the population voices like Ken Smail, Lindsey Grant, William Stanton, Albert Bartlett, and Andrew Ferguson of the Optimum Population Trust (OPT) in the UK (**www.optimumpopulation.com**) agree that about two billion people is the maximum population (carrying capacity) the earth can support with the sustainable incoming solar energy base, but not the previously stored, concentrated, bio-energy in the form of fossil fuels.

What should we do as we enter the critical year of 2006?

The Survival Plan

Read again The Last Chance for Sustainability on page xv. Re-emphasize point number 6 "reduce population." If we start now to reduce energy consumption, population, and eco-devastation, there may be time to stretch (mitigate) the transition over decades and several generations.

Rather than wait a little longer and squeeze every last bit of fossil energy out of the earth as quickly as possible, we should make an about-face to begin a planned descent. This will make the year-to-year adjustments we must make more gradual and less drastic. Americans cannot dictate to the rest of world. We can only set an example.

The Depletion Protocol proposed by ASPO (The Association for Study of Peak Oil and Gas) is an attempt to convince world leaders to agree to reduce consumption to match world depletion rate. The intent would be to reduce demand of individual countries to balance with the declining supply. This would be a form of rationing to insure stabilization of prices and allow poorer countries and individual consumers to afford their needs. Richard Heinberg is taking the lead in 2006 to bring the Depletion Protocol to the U.S. by establishing an office and publishing a book.

By reducing consumption proportional to depletion, preferably by rationing instead of taxation or market price inflation, the gains of efficiency improvements will not be lost to continued growth. Unfortunately, even the oil Depletion Protocol (or the similar Five Percent Plan proposed in Chapter 5 of this book) will not suffice if population is not concurrently reduced.

This is the crux of this 3rd edition. Population must decline as fossil fuels run out. These parallel downward trends will make it possible to reduce the devastating impact that humankind and industrialization is having on our delicate eco-system. This assumes that the declining population will not pillage the earth's remaining resources as has been typical of earlier societies in their death throes.

What more can be said. We have even less time to get started than a year ago. But, we can't give up.

Hopefully the sharp increases in energy costs will wake up the main-stream public led by the number one consumers in the world ... ourselves. Those who argue against the impending triple-crisis of civilization: energy, population, and eco-devastation, are flying in the face of math, science, and time ... time lost that will seal our collective fate.

Chapter 1

Energy-Crisis Overview–2003

We are on a collision course with disaster. In the last four generations (100 years), we have built a very tall house of cards and enjoyed a party provided by plentiful, low-cost fossil fuel. We've developed the technology to convert plentiful energy into an easy lifestyle. This has been going on just long enough that few in the developed nations remember what life was like before this luxury. It took hundreds of millions of years for the earth to accumulate various forms and small pockets of concentrated fuel. Now, we are using it at such a prodigious rate that at least the liquid and gaseous components will become scarce in the next decade. The age of stored fossil-fuel energy will have lasted, in total, only about 200 years. A similar life-altering asteroid might impact the earth every few hundred million years. Our man-made crisis is at our doorstep now. Rather than denying this, leaving our survival to others, or just giving up, read on. There still may be a last chance to achieve a sustainable civilization. It is up to us as intelligent individuals acting in concert to identify and follow that path. We are the problem. We are the only hope for the solution. Time is critical. Every moment wasted means less chance for survival. In the last two years we may have lost our best chance for a safe landing.

For thousands of generations our ancestors subsisted in a delicate, harsh balance with the environment. Like all organisms

1

our bodies have the ability to produce at least enough energy to procure food (fuel) which, in turn, provides our personal energy along with a little left over to fight with competitors for resources or to procreate and survive lean times. In good years, human population increased because the population of any species reflects the abundance of the environment. As soon as the resources dwindled, either because of resource depletion or climate change, the population declined. This forced the population to remain stable for tens of thousands of years.

Since the rapid advance of technology in the last century, which enabled us to utilize previously stored highly concentrated energy, world population has exploded at an exponential rate from one billion people to over six billion.

Any dynamic increase like population growth or resource consumption cannot last. The greater the growth or movement in one direction, the more severe will be the ultimate correction. What goes up, must come down.

Some changes are imperceptible in our human time frame. It takes recorded history, archeological research, and science to accurately understand the past, assess the facts, and predict the future. To deny these natural and scientific observations will be our demise.

Less intelligent species can be forgiven for not seeing or reacting to the external forces which control population and resource consumption. Humans, on the other hand, are gifted with magnificent brainpower and the ability to learn from the past and plan for the future. Yet, we are blind if we don't use our extreme intelligence to anticipate and hopefully cushion our impending collision with the coming energy and closely related population crises.

There are many voices in the wilderness warning of the impending crash, but for all practical purposes nothing is happening. We are just shrinking the time frame to our demise. Later in this book I discuss further the movement of environmentalism. Why we don't want to hear or act on these messages is varied.

One author, Peter Seidel, in his book, *Invisible Walls*,[1] has directly addressed this subject. Time is rapidly running out. We will soon lose any opportunity to act appropriately while we do nothing but let short-term interests determine our path.

If we lose the advanced technology we are presently enjoying, no future civilization will ever again be able to achieve a similar level of freedom from human labor. The fossil fuels and their high plateau of support of our lifestyle will be forever lost.

The necessary downsizing of consumption will require many changes. These need to be understood and made by all, not just a few. The only hope for survival on our life raft is for leadership and peer pressure to insure that everyone shares the commitment and a few don't rob the provisions. Fortunately, much of the Industrialized World is ruled by democracies with elected leaders who should answer to informed constituents and not to inanimate corporate structures concerned primarily with short-term profits regardless of the impact on natural resources.

In the 2004 critical election year, we in the U.S. must support candidates most concerned with the immediate impact of energy on the survival of humankind.

Unfortunately by 2005 when this book is being reprinted, the 2004 election is behind us and neither party focused on the pending fossil-fuel crisis as a fundamental issue. As usual, both candidates reflected an uninformed constituancy that does not know of or is being mislead about energy issues. The proposed energy bill does not focus on fossil-fuel depletion. It only gives lip service to conservation and encourages us to keep going as we are with even more intensive drilling until all is gone. Then what?

🌲 🌲 🌲 🌲 🌲

Our present complacency is encouraged by a mix of the following (not essentially in order):

- **The crisis is not urgent. Leave it to future generations.** Wrong! The crisis is here now. In the next 5 years, major changes will occur. The direction of these changes depends on us. Our children will be left with nothing but scraps if we don't act now. The sooner we start to reduce our fossil-fuel consumption, the more hope we have. We are stealing the future from everyone.

- **Religion and prayer will save us.** Wrong! God is sending us a message now. We should heed it well. The free-energy party is man-made including all the luxuries we take for granted— cheap and plentiful food, plush and easy transportation, warm houses and hot baths. We cannot even return to the harsh existence of the Middle Ages. There are too many people now, and in the Industrialized World, we have lost our subsistence skills. If heaven and afterlife are the salvation for some, we are not leaving enough finite resources for our earthly descendants.

  The religious community may in fact be a ray of hope in its recognition of the energy crisis and call to action as stewards of the earth. There is a groundswell of activity by Christians, Muslims, and Jews joining with other beliefs in a common concern for environmentalism and the subject of finite-fuel sources. Prompt dissemination of the energy-crisis message may be facilitated through religious networks. For those with strong biblical beliefs, Noah planned ahead and was extremely proactive when he was warned of the coming flood. (See Chapter 7 for further discussion of religion and energy.)

- **The scientists will save us.** Wrong! After scientists taught us how to use essentially free, previously-stored fossil energy, they are now telling us we are running out. Unfortunately, the public is not hearing this message, possibly because it is in complex books and technical language. An exception is the excellent book, *Hubbert's Peak* by Kenneth Deffeyes[2] (the follow-up book by the same author is, *Beyond Oil: The View From*

Hubbert's Peak). Other books are reviewed in the new Chapter 2. These recent books explain in clear, undeniable detail the history and consumption of petroleum as our most important fossil fuel. It makes little difference if we have 20 or 40 years left. The truth is, we can't go on this way. The more we have left and the sooner we start to conserve, the better our chance for survival.

One other important point to keep in mind is that any possibility for a significant, high-technology future requires the springboard of a functioning, energy-intensive society such as we have today. The infrastructure and investment required for controlled nuclear fusion or other energy related breakthroughs cannot come from the ashes of anarchy or a failed civilization.

• **There is a strong environmental movement growing in the world today.** Unfortunately, the messages are soft, sometimes misdirected to other priorities and may lead to false complacency. Other weaknesses of well-meaning, non-profit advocacy groups (sometimes referred to as non-government organizations or NGO's) are:

1. They churn around and around preaching to the choir, their own members. With memberships of up to a million, each is still less than one percent of the U.S. population. They are not reaching enough of the masses to be effective. Glossy periodicals and tiny environmental actions delude the public into thinking there is progress. Instead we are losing ground, especially with leadership (right or left) concerned with re-election based on optimism, near-term gratification, and, where deemed necessary, military intervention to maintain energy resource flow with little or no planning for future supplies and resources.

2. Even the closely related subject of climate change is not making an impact since the time frame is measured in hundreds of years not decades. Not many people care. The public doesn't feel an urgent crisis. They may think global warming would require less energy for heating (or

plowing snow) and be a good thing. Global warming will become a more serious problem after we run short of oil and natural gas as we obtain more and more of our energy from dirty fossil fuels like high-sulfur coal and difficult to extract, non-conventional sources like tar sands and heavy oil. By that time, in a few years, our civilization will already be in serious trouble if we do not address the immediate problem of dwindling cleaner fuels. The few extra years we gain by using environmental unfriendly dirty fuels, like tar sands and coal, will not avert the coming crisis.

3. Other issues like species extinction, pollution, water quality, public health, etc. are not of urgent interest to most Americans. **We need to focus on one message—we are running out of fossil fuels.** The tank is moving toward empty. Our energy-guzzling economy is hurtling down the road with no hope for refueling. The truth is, if we start to immediately address the fossil-fuel depletion problem, many of the other related environmental problems will become less serious. Even if we continue on our present path, all fossil fuel and closely-related population problems will eventually resolve themselves through the harsh laws of depleted resources. Mother Nature will triumph in the end. As just another living species, our remaining descendents will be cruelly forced to come to long-term sustainable equilibrium with the environment and remaining resources.

Sustainability is the environmental buzzword for a steady state process. To sustain means to maintain, support, and keep alive. For our purposes and for the future of civilization, we need to find a balance between energy intake and energy expended. This is equally true of our own bodies. Each of us is a complex machine able to convert energy (measured in Calories) from the food we eat into work. For eons, our ancestors survived by expending nearly all their energy just to grow or hunt for enough food. In today's age of easy energy and resulting plentiful food, we each can eat far more Calories than we need. This leads to obesity and other related ills.

The energy party the Industrialized World is now enjoying helps to explain many disparities and conflicts. Energy is the difference between modern military might and legions of foot soldiers. The conspicuous consumption of the Western World, now visible to the Third World via modern media, leads to great animosity. Why do they hate us? Our gross consumption of energy, much of it coming from Third World resources, is visible and alarming to the rest of the world. Modern Western Europeans question American squandering of the world's energy resources, while at the same time they are heavily taxed on fuel for transportation.

This book is an engineer's attempt to envision a model of a low-energy, sustainable, modern lifestyle along with a timely plan to get there. No attempt is made to alienate specific groups who may be liberal or conservative, religious fundamentalists or non-believers. We are all on one lifeboat and must share the resources and pull in unison if there is to be any future worth living. All of us will have to make the necessary changes. These will not necessarily be sacrifices since in many cases we will be better off.

Basic science and math clearly show that we must make significant reductions in our energy consumption. Our only hope is to start immediately while we still may have enough fossil energy left to make a controlled transition.

Chapter 2

A Two-Year Update 2005

By 2005, the energy picture in the U.S. is essentially unchanged. Life goes on with business as usual, except for fifty percent higher fuel prices and another 55 billion barrels of oil used in these two years, gone forever. The American economy rolls along on still relatively cheap fuel and projected growth. A most wrenching experience for this writer is to research, read, and write about our coming energy crisis, then go out into the "real world" (Which is real?) and join the party, all the time knowing that we're living in a surreal dream world. We're moving into the second half of our short (several hundred years) fossil-fuel epoch, and yet we're still having a grand fiesta seeing how fast we can use it up.

The SUV in a "feel good" ad with the successful parents and smiling toddlers is a symbol of why there can be no future for those very kids. We are throwing away their energy endowment. This is a bitter pill to swallow, but if we don't first recognize our terminal illness, how can we possibly avoid the consequences?

It is no longer possible to read, keep up with, or credit all the numerous "peak oil" websites that appear daily. I believe the Internet goes unnoticed by the typical American driving merrily along while lulled into believing by leadership and media, that all is well.

One example of typical web information is by Steven Lagavulin (**www.deconsumption.typepad.com**) dated March 16, 2005 and titled "The Most Important Thing You Don't Want to Know About Peak Oil." To quote:

> *The "oil grab" is in fact already on . . . a desperate resource war is emergent . . . but the true gravity of the situation is only scarcely beginning to come to light. The markets have already accepted the long-term "bull market" in oil prices due to increasing demand. What they don't accept yet (or understand) is the mounting supply problem. I believe that the only hope of changing things is by building a consensus among people. This needs to happen very quickly . . . .*

Almost all the energy we have, had in the past, or presently, continues to flow from the sun onto the surface of the earth. This radiation energy combines the carbon in the short gaseous molecule of carbon dioxide (CO_2) with the hydrogen from water (H_2O) through a process called photosynthesis. Familiar green plant life forms from algae and moss to large trees have been taking in this energy from the sun for millions of years and using it to make larger molecules of hydrogen and carbon commonly referred to as carbohydrates.

Under unique conditions of heat and pressure, some of these carbohydrates were compressed into tightly-packed forms of energy like oil and coal, or trapped in pockets of the earth as natural gas. The incredibly long time this process took allowed the earth to be endowed with sporatic sources of concentrated stored energy called fossil hydrocarbons or fossil energy. Our Industrial Age and concurrent 6-fold (one billion to 6 billion) population explosion in the last 150 years has only been possible because of the extraction and utilization of this previously stored energy.

There are other less concentrated forms of energy including the very dilute continuous inflow of sunshine. These will be discussed later in this chapter. The important point is that modern, industrialized civilization has evolved only because of the availability of an incredible amount of energy. Today, about 86% of this energy comes from finite deposits of previously stored hydrocarbons, which are now rapidly being depleted.

The exceptions to the energy, which came from or is now coming directly from the sun, are moon gravitational pull (tides), geothermal energy (from deep in the earth), and nuclear energy.

Since all forms of energy can do work, they can be converted from one form to a different secondary form, but not without some loss in the conversion. The ratio of the secondary energy to its earlier form is called efficiency. The conversion may be very efficient or very poor depending on the procedure used.

As should be clear by now, energy is the underlying theme of this book. To grasp the complete message, it helps to have a basic understanding of this complex subject. There are many textbooks available but few are easy to understand. An exception in layman's language is entitled *Energies* by Vaclav Smil.[1] His latest book, *Energy At The Crossroads*, and other new books, which have been published in the last year, are reviewed later in this Chapter 2 update.

For further background on the fundamentals of energy, the reader is referred to Appendix 6, "Energy 101 ... Energy, Work, Power." It is not necessary to understand the math and technicalities of energy to grasp the message of this book. This section as been moved to the appendix for readers who desire a better understanding of the terms, units, and numbers. In simpler terms, I will attempt to provide enough background to better understand my book and help sort out all the conflicting inputs we hear today. We need at least a numerical appreciation for the huge difference in energy content between different sources. Numbers are also necessary to compare the magnitude of concentrated fossil energy with the relatively puny human, animal, and continuous inflowing solar-energy forms. Understandable equivalent units are also necessary to appreciate the original finite endowments of stored hydrocarbon energy and to compare these with usage rates and remaining reserves.

Confusion concerning energy is a primary cause of poor decision-making and the lack of urgency that is so prevalent today. Numbers are also necessary if we are to intelligently search for a sustainable future and a realistic plan to get there within an acceptable time frame.

EIA (ENERGY INFORMATION AGENCY) UPDATE OF ENERGY SOURCES

The following table shows U.S. consumption from all energy sources in 2003 and 2004 in equivalent billion barrels of oil (EBBO) from the EIA tables 1.3 and A.1 and Annual Energy Outlook 2005, **www.eia.doe.gov/energyupdate**:

Fossil	2003	2004	% of Total
Oil	6.2	6.35	40.0
Natural Gas	3.67	3.66	23.0
Coal	3.61	3.61	22.7
Subtotal, fossil	**13.49**	**13.62**	**85.7**
Non-Fossil			
Nuclear	1.27	1.27	8.0
Hydro	0.45	0.45	2.83
Wood	0.332	0.336	2.09
Biowaste	0.084	0.088	0.554
Ethanol	0.038	0.047	0.296
Geothermal	0.054	0.054	0.34
Solar	0.010	0.010	0.063
Wind	0.018	0.023	0.145
Subtotal, non-fossil	**2.242**	**2.278**	**14.3**
TOTAL	**15.74**	**15.90**	**100%**

Oil still provides forty percent of U.S. total energy with about one-half of that being used for transportation gasoline alone. To repeat, twenty percent of our total energy, which is one-eighth of the world oil consumption, allows us to indiscriminately drive huge personal vehicles whenever, wherever. Total world-oil consumption is a billion barrels every twelve days. The U.S. uses about one-half of the world's gasoline at a rate close to 400 million gallons (or one and one-half gallons per person) **each day**.

Natural gas and coal each provide twenty-three percent of our energy. Together, they provide the bulk of our non-transportation heating industrial and electrical fuels. Combined with oil, the three finite fossil fuels provide eighty-six percent of our total energy—a proportion which is also true of the world as a whole, led by the U.S., lumbering along like an aging dinosaur, unaware of its pending doom in the next few years.

As always at this point of discussion, we come face to face with the extreme and absurd dichotomy between future realism and wishful thinking extrapolated from past years. The EIA projects in their very comprehensive twenty-eight page, twenty-year forecast that the petroleum party will go on at least until 2025. For instance, they project forward from 2003:

- Vehicle miles traveled will increase (56%) from 2602 billion miles to 4053 billion miles.

- Air-seat miles will increase (63%) from 932 million miles to 1520 billion miles.

- Freight-oil usage will increase (60%) from 2.13 million barrels per day to 3.4 million barrels per day.

- Residential-electricity usage will increase (43%) from 1267 billion kilowatt hours to 1810 billion kilowatt hours.

And on and on! What happens after 2025? The EIA's answer is that they don't forecast beyond 2025 (per my personal communication with their office).

During this same twenty-two year period the EIA projects that:

- U.S. domestic crude oil (Alaska plus lower 48) production will decline slightly from 5.74 million barrels per day to 4.73 million barrels per day. In the twenty-two year period from 2003 to 2025, this EIA forecast production would total about 44 billion barrels with 17 billion barrels still remaining. The Association for the Study of Peak Oil and Gas (ASPO) estimates present U.S. petroleum reserves in 2003 totaled only 20 billion barrels, and then we're out! (See Appendix 5 in this book.) This is a difference of (44 plus 17) or about 60 billion barrels vs. 20 billion barrels. Amazing math.

- World petroleum prices will increase (in 2003 dollars) from $24.10 per barrel to $30.31 by 2025. Whoa! We already went to $55.00 per barrel in October of 2004 and again in March & June 2005. How can we believe in a nation's future growth, oil production, and oil consumption based on comprehensive forecasts, which are already dead wrong by a factor of two?

I don't want to belabor the huge discrepancy between what oil experts throughout the world are telling us and what our EIA wing of the DOE (Department of Energy) is forecasting, but this huge difference is a focal point of everything we must do to survive, starting now ... immediately!

For backup references, I refer the reader to a multitude of books and websites, updated in this new chapter and Chapter 3.

Some typical highlights of current world peak oil news are quoted from the ASPO newsletter, number 51, March 2005, **www.asponews.org**:

(Item 489) "Oil Debate Revving Up"
This article from The Australian speaks of a new awareness of the oil depletion issue (Niel McDonald, 03 February 2005).

(Item 496) The IV International Workshop on oil and gas depletion will be held in Lisbon, Portugal in May 2005. Contributions from 25 international leading specialists include:
• Reality in oil exporting countries—the supply limits
• Impacts of depletion in oil importing countries the demand pressure
• How much regular oil and non-conventional oil—Utopia versus reality
• The case for political action—the depletion protocol
• The world past peak oil age

For the complete conference proceedings in detail, see **www.cge.uevora.pt/aspo2005**

(Item 497) "Entering the Age of Oil Depletion"
This is a U.K. conference in Endinburgh, April 25, 2005. Speakers and topics include:
• Colin Campbell—"The End of the First Half of the Age of Oil"
• Matthew Simmons—"Can Market Efficiency Overcome Depletion? Of Why Economists Don't Get It"
• David Spavin—"Transport–An Oil Crisis and More"

(Item 499) "The G7 Ministers Begin to Grasp Peak Oil"
"Saudis consider knowledge of their reserves as a source of power, which might be an oblique way of describing loss of power if their reserves are in reality less than widely supposed" (London, Feb. 4, 2005).

(Item 500) "Venezuela's New Ties: Russia and China" . . . quoting World Watch **www.energyintel.com** . . . "President Hugo Chavez can count Russia among his new friends as he talks of steering Venezuela's oil future away from its traditional

15

dependency on the U.S. . . . . Chavez's visit to Moscow at the end of last year has brought quick results . . . . "

(Item 501) The IEA (International Energy Agency, consistently optimistic, like our EIA) deserves praise on its limitless planet.

. . . oil resources were near limitless, with production being simply a matter of investment, technology and politics . . . its statements and fatuous business-as-usual scenarios, reflecting outdated economic theory, must have a political motive to spare member governments from the uncomfortable challenge of facing the reality of oil depletion, with all that it implies.

(Item 504) The U.S. Department of Energy addresses peak oil (Hirsch, R.L., et al, "Peaking of World Oil Production. Impacts, Mitigation, and Risk Management," DOE NETL, February 2005). This report clearly acknowledges the "peaking of world production." Mitigation options based on "crash program implementation" are presented because "mitigation cannot avert massive shortages unless it is initiated well in advance of peaking" and "world oil peaking represents problems like no other. The political, economic, and social stakes are enormous. Prudent risk management urgent attention and early action."

I have this comprehensive, 91-page report in its entirety. It is thoroughly researched and includes 149 footnotes. Although it does not imply an endorsement by the United States government, it reads as thoroughly and with as much pessimism as any of the recent books on fossil-energy depletion. It carefully addresses the seriousness of natural gas scarcity as well as the economic implications of the crisis (in addition to peak oil) now facing the U.S. and the world. The solutions for "mitigation" begin, like my book, with extensive conservation measures with automobiles and light trucks, "the largest targets for consumption reduction." This report should be on the desk of every legislator and governor, but lately it appears to have disappeared from Internet access (**www.netl.doe.gov**). The synopsis is still available at **www.asponews.org**, number 51.

(Item 505) Decline in the deep water Gulf of Mexico "discoveries have declined from 21 in 2002, and 17 in 2003, to 12 in 2004, despite every effort and impressive technological achievements."

(Item 506) Indonesia contemplates leaving OPEC "as its production is in terminal decline, having passed midpoint in 1992." (Yet, Indonesia supplied the U.S. with an average of 28 thousand barrels a day during 2003.)

(Item 507) Heresy is alive and well, "a group of prestigious U.S. foundations have sponsored a major flawed study . . . oblivious of depletion and material resource limits."

I have downloaded this bipartisan study from **www.energycomission.org**. This report, issued December 8, 2004, attempts to guide national energy policy based on "incentives to spur global-oil production, to increase domestic-vehicle fuel economy, and to increase investment in automotive fuels." The report gives many good recommendations but ends agreeing with the EIA that the "U.S. will consume 43% more oil and emit 42% more greenhouse gas emissions by 2025."

(Item 508) Recognition of Peak
"Bloomberg's news of February 18, 2005 has assembled a number of oil company statements that directly and indirectly speak of peak. The most direct are the words of the famous corporate raider, Boon Pickens, who says, 'We're damn close'."

(Item 510) Exporting birthright
"US Department of Energy reports the origin of the country's 4.4gb (billion barrels of oil) of annual imports. The two largest suppliers are its neighbors, which have lost a degree of sovereignty under NAFTA, and are more or less forced to deliver. In fourth place is Venezuela, which seeks new customers (see item 500). Most of the others are major exporters, but what stands out as remarkable is the presence of the United Kingdom which itself becomes a growing new importer from 2006 onwards. Since American companies own rights to many of the North Sea fields, they are of course at liberty under the present regime to export to their home country, but it is strange for a country to feed some-

one else while starving itself of this vital ingredient for its economic survival."

At this point, I will depart from the ASPO newsletter, number 51, March 2005 but continue the last train of thought by listing eleven major sources of U.S. petroleum (from **www.eia.doe.gov**) with observations and several recently cited references regarding the future reliability of these sources. Again, as is true throughout my book, EBBO refers to equivalent billion barrels of oil. These numbers are also exclusive of natural gas plant liquids (NGL).

United States—1.91 EBBO (way past peak)
The major source of our oil is still our own country, although we clearly went over peak in 1972 and are now squeezing out the last ten percent (20 EBBO) of our original endowment of around 200 EBBO. We were the Saudi Arabia of oil in the 20th century; our tank is almost empty.

A closer examination of the voluminous EIA Annual Energy Outlook 2005 reveals the disturbing fact that in 2003, we EXPORTED 0.34 billion barrels of our remaining precious oil. Will we continue to export our children's inheritance to the highest world bidder? Will we export ANWR oil? Just this 0.34 EBBO is the same order of magnitude as the solar energy achieved by putting 4 Kw of photovoltaics in 40 million homes in 40 years. (See Chapter 4.)

Canada—0.57 EBBO (past peak)
Our NAFTA neighbor has about 6 EBBO left of its original 25 EBBO endowment or regular oil. This does not include tar sands; more about that later. Regular oil is clearly way past peak.

Mexico—0.57 EBBO (past peak)

Our other NAFTA neighbor has about 25 EBBO left of its original 55 EBBO endowment . . . better shape than Canada but no tar sands. They are at about peak and heading for decline. Typical of recent releases regarding Mexican oil, the following by Thomas Black appeared in Bloomberg news on February 24, 2005:

Exxon-Mobile, BP, and other international oil problems say they will shun Mexico's offer to bid on contracts to drill for oil in the Gulf of Mexico because the proposal isn't profitable enough. The contracts 'do not recognize the high cost and high risk of deep water and so really don't provide us with the ability to do that', said Tim Cejka, president of Irving – based Exxon-Mobil's exploration unit.

Saudi Arabia—0.63 EBBO (before peak)

Our primary OPEC trading partner claims to have 200 EBBO left of their original 300 EBBO. This sounds more encouraging except Saudi production is earmarked to keep pace with increasing world demand concurrent with making up for lost production in the many declining countries. Also their reserve estimates are not transparent and may be optimistic. For instance, the following article by Steve Everly appeared in the *Kansas City Star* on February 16, 2005, "Meeting U.S. projections for increased Saudi Arabian oil production by 2025 would be a 'difficult battle,' the head of Saudi Arabia's oil company said Tuesday . . . Abdullah Jum'ah, chief executive of the Saudi Arabian Oil Co., said the company had developed scenarios to increase production to 15 million barrels but not to 22.5 million barrels a day . . . Saudi Arabia's role in the world oil picture is crucial. It has 25 percent of the world's reserves and is the largest net oil exporter. It supplied the United States 1.5 million barrels of oil a day in 2004 . . . 15 percent of the country's oil imports." On April 26, 2005, the Saudis, in order to bring world prices down, advised President Bush they will increase their output by 30% (about 4% of world consumption) by 2009. We shall see. More promises?

Another clue to the precarious nature of Saudi reserves and production capabilities can be inferred from the Bank of Montreal analyst, Don Coxe. Working from their Chicago office,

Coxe states, "On April 12, 2005 the combination of the news that there's no new Saudi Light coming on stream for the next seven years plus the 27% projected decline from existing fields means Hubbert's Peak has arrived in Saudi Arabia." Coxe goes on to say, "as if that weren't bad enough news, the Saudis claim they need at least $32 a barrel to justify new production, because ... new production ... requires water flooding. Water flooding on newborn Saudi wells? Isn't water flooding the viagra of aging men?" (**www.english.aljazera.net**) The most revealing insight to the truth about Saudi oil is the recently published book by Matthew Simmons, *Twilight in the Desert*. See review later in this chapter.

Venezuela—0.43 EBBO (at peak)

Our not-so-friendly South American neighbor with one-half of its original endowment of 95 EBBO remaining, Venezuela is at or near peak of regular oil. "Venezualan president Hugo Chavez, whose country is the world's fifth largest oil exporter, said that the world should get used to high oil prices. 'The world should forget about cheap oil . . . it won't happen,' Chavez told a news conference in the Indian capital Saturday . . . added later however, he was certain the prices 'will go up even further' than the 40 dollar to 50 dollar range." Chavez comments came after India, which imports 70 percent of its crude oil needs, signed a deal to take a 49 percent stake in Venezuela's San Cristubal oil field" (Agonce France–Presse, March 5, 2005, **www.uk.news.yahoo.com**). In December 2004, President Chavez met with his Chinese counterpart, Hu Jintau, in Beijing to discuss a bilateral agreement regarding access to Venezuela's energy market. In Chavez's words, "This is what is needed in the world in order to break with unilateralism." This meeting was preceded by Chavez's renewed calls for the creation of PetroSur, a Latin American version of OPEC (from **www.canadiandemocraticmovement.ca**).

On April 26, 2005, Chavez, "a close friend and ally of Cuban leader Fidel Castro ... decided to begin selling 53,000 barrels of crude a day to oil-import-dependent Cuba under preferential terms" (**www.forbes.com/home/foods**).

Nigeria—0.304 EBBO (before peak)

Nigeria has about 33 of its original 55 EBBO remaining and has not yet peaked. This country is typical of First World extraction of African Third World resources with little concern for benefits flowing to the indigenous population.

Iraq—0.176 EBBO (way before peak)

With 100 EBBO of its original endowment of 132 EBBO remaining, this country is clearly the wild card. No small wonder the world has focused its attention there. This is why we desire friends in charge of this country instead of leaders or insurgents, who would like us to leave. The fact remains, at 1.9 million barrels per day, Iraq oil production is still below the pre-war level of 2.5 MB/D.

Angola—0.132 EBBO (at peak)

With about 5 EBBO of its original endowment of 10 EBBO, this country is about at its peak and shares the same sorrows as Nigeria.

United Kingdom (North Sea)—0.132 EBBO (past peak)

(See comments under ASPO, item 510) The U.K. has already used about 20 EBBO of its earlier original endowment of 32 EBBO. This area is clearly in decline and is not expected to meet its own needs beginning in 2006. How can it be expected to fuel the American life style? An additional reference can be found in an Associated Press release dated February 25, 2005 (**www.forbes.com**). "Statoil ASA on Friday presented a plan for squeezing more oil and natural gas out of the aging Statfjord Development, the largest field in the North Sea. Statfjord, one of the oldest Norwegian fields, was discovered in 1974 and began producing in 1979. It straddles the North Sea median line between Norway and Britain."

Algeria—0.041 EBBO (before peak)

This country has recently become politically friendly to the U.S. with 16 EBBO of its original 28 EBBO endowment remaining (plus much natural gas, which is already earmarked for LNG or liquid natural gas markets in Europe and the Far East).

Kuwait—0.04 EBBO (before peak)

Like its neighboring Middle East countries, Kuwait is not yet at peak. It still has about 60 EBBO of its original 90 EBBO endowment remaining. It is most strategically and politically friendly to U.S. but supplies a large part of the European market.

To conclude, the U.S., with an annual consumption of about 6 billion barrels of oil is clearly relying on a number of major sources (starting with itself), which are well beyond or near the peak of the production. This fact increases pressure on the remaining oil producers to make up for the increasing shortfall as well as to accommodate the fast-developing nations of China and India. All the world oil producers can barely keep pace with the present total consumption of 30 billion barrels annually (about 82 million barrels per day) of which about one-fourth is being consumed by the U.S. There is no longer enough excess capacity to accommodate additional growth. Economics 101 tells us that the commodity price will now rise as tightening supply prohibits additional growth in spite of increased demand. At some point in the near future, supply will start to decline regardless of price because the dwindling commodity (world oil) is in permanent decline. The U.S. energy bill now being debated (June 2005) in Congress will do nothing to allay the specter of peak oil. It is quite obvious that our elected leaders do not fully understand the dilemma we're in as they debate the nuances of hydrogen, biofuels, renewables, ANWR, auto fuel-consumption standards, and avoidance by big business of MTBE/drinking-water lawsuits.

As I started to update this dilemma in the middle of March 2005, rural Maine was buried under almost three feet of snow. I can think of no way, 25 years from now, that we will push snow around like we do now with gasoline and diesel fuel. Every major northeast storm requires untold millions of gallons of oil energy. This cannot be done with electricity, coal, nuclear, solar, whatever, except possible biofuel, which is our personal-food energy. The same concept holds true for nearly all transportation energy. Only petroleum made possible the industrialized and globalized world we know today. By April, the snow is gone. The spring buds and birds give promise that Mother Nature will surely survive after our fossil-fuel party is over and Homosapiens have returned to the harsh reality of some form of long-term sustainability.

We may have already passed the point where it is possible to recognize and correct our terminal fossil-energy illness. If our doctor missed the opportunity to diagnose and prevent a terminal illness, he/she would be accused of malpractice. For years, our leaders (right and left, except for Jimmy Carter in the early 70's) have clearly failed us. Is this political malpractice?

OTHER NEW INPUT

Since starting this project two years ago, awareness and concern about fossil-energy depletion has exploded. A small part of the American public is waking up largely due to the fifty-percent increase in the cost of gasoline and heating oil. Even national mainstream media (in addition to many daily newspaper articles across the country) have ventured into "peak oil." For instance:

- "The End of Cheap Oil," *National Geographic*, June 2004 and another released cover story, "Global Warming, Bulletins from a Warmer World," *National Geographic*, September 2004.

- "Gas Guzzlers Shock Therapy," Jane Bryant Quinn, *Newsweek*, August 16, 2004.

- "Over a Barrel," Paul Roberts, *Mother Jones*, Nov/Dec 2004. "Experts say we're about to run out of oil, but we're nowhere near having another technology ready to take its place."

- A synopsis, "Limits to Growth, the 30-Year Update," Meadows, Randers, Meadows, *The Reporter*, Spring 2005.

Many new books have been written. Those I have read are listed below with my comments:

The End of Oil: On the Edge of a Perilous New World by Paul Roberts, Houghton Mifflen, 2004 (389 pages). This is a comprehensive, basic primer that focuses on the oil component of our fossil-fuel dilemma. This book is written by a professional who knows the whole story and how to present it to a mass audience. Paul Roberts has also written numerous articles such as the one mentioned above in *The New Yorker* and *Mother Jones*.

Powerdown: Options and Actions for a Post-Carbon World by Richard Heinberg, New Society Publishers, 2004 (209 pages). Richard Heinberg, in my estimation, has assumed the mantle of the U.S. leader in the peak-oil movement. His first book on the subject, *The Party's Over*, has become the standard reference U.S. work on peak oil. This second book follows the first with more current information and attention to a post-peak society. Heinberg lectures extensively on peak oil and was scheduled to talk at the worldwide ASPO conference in Portugal, May 2005. His monthly *MuseLetter* has become (along with ASPO News) the basic monthly update on a fast-evolving crisis. We all thank you, Richard, for your wisdom, guidance, and eloquent presentations (see **www.museletter.com**).

The End of the Oil Age by Dale Allen Pfeiffer, self-published, 2004 (264 pages). Dale Allen Pfeiffer is also a recognized leader in the peak-oil movement who was raising alarm bells before most of us were concerned. He personally guided Michael Ruppert (**www.fromthewilderness.com**) into the subject of peak oil. This helps explain the wide variety of related and troubling subjects that the *From The Wilderness* website covers. Pfeiffer can be reached at **www.lulu.com**.

Beyond Oil: The View from Hubbert's Peak by Kenneth Deffeyes, Hill and Wang, 2005 (202 pages). This book is Deffeyes' long-awaited sequel to Hubbert's Peak. As a career energy geologist and Princeton professor, Deffeyes adds the most informative, sometimes witty, and comprehensive background to all aspects of fossil fuels.

Collapse: How Societies Choose to Fail or Succeed by Jared Diamond, The Penguin Group, 2005 (573 pages). This is an excellent treatise by a respected, Pulitzer Prize winning author. My only reservation is that this book deals more with climate change and all resource depletion and does not focus on fossil energy. The message in this book also infers possible solutions, which may not be realistic after fossil-fuel depletion (see **www.museletter.com**, number 155 for a comprehensive review *Collapse*).

High Noon for Natural Gas: The New Energy Crisis by Julian Darley, Chelsea Green Publishing, 2004 (266 pages). This is the definitive work on the crisis situation of the natural gas component in the fossil-fuel trio (oil, natural gas, and coal). Chapter headings like "Where on Earth Are We Now?"; "Where Are We Going?"; and "But What Else Can We Do?", give indications of this book's straight-forward, comprehensive coverage. *High Noon for Natural Gas*, plus a companion video, *The End of Suburbia*, are must references for anybody (which should be every American) interested in the peak-oil and the fossil-energy crisis. Both can be ordered from the number one U.S. source and website **www.postcarbon.org**.

The Long Emergency: Surviving the Converging Catastrophes of the Twenty-First Century by James Howard Kunstler, Atlantic Monthly Press, 2005 (305 pages). This new book offers realistic and eye-opening visions of a post-carbon world. Like *The Oil Age is Over* listed next, the situation (present and future) is not pretty. There are no easy solutions. For example, the following quotes from Chapter 1, "Sleepwalking Into The Future," set the tone of the whole book:

> *As the industrial story ends, the greater saga of mankind will move into a new episode, The Long Emergency. This is perhaps a self evident point but throughout history, even the most important and self-evident trends are often completely ignored because the changes they foreshadow are simply unthinkable. That process is sometimes referred to as an 'outside context problem,' something so far beyond the ordinary*

experience of those dwelling in a certain time and place that they cannot make sense of available information. The collective mental static preventing comprehension is also sometimes referred to as 'cognitive dissonance,' a term borrowed from developmental psychology. It helps explain why the American public has been sleepwalking into the future.

The Long Emergency is going to be a tremendous trauma for the human race. It is likely to entail political turbulence every bit as extreme as the economic conditions that prompt it. We will not believe that this is happening to us, that two hundred years of modernity can be brought to its knees by a worldwide power shortage.

The Oil Age is Over by Matt Savinar, self-published, 2004 (181 pages). This brilliant young lawyer in his mid-20's has written this most disturbing book in a very creative question and answer format. Matt just passed his California bar exam and was "excited about a potentially long and prosperous career in the legal profession . . . since learning about peak oil, those dreams have been radically altered . . . the information in this book is not for the faint of heart or the easily disturbed." Matt's book is most thoroughly researched (280 endnotes and sources). It is easy to pick up and start reading at any point, but it is not for peaceful sleeping afterwards. See **www.lifeaftertheoilcrash.net** for more information and ordering details.

The Collapsing Bubble: Growth and Fossil Energy by Lindsey Grant, Seven Locks Press, 2005 (74 pages). This short consise book is the best of the few which combine the critical challanges of imminent fossil fuel depletion with the need to reduce population to much lower levels. This is the best new book on the triple-crisis facing civilization.

A Green History of the World: The Collapse of Great Civilizations by Clive Ponting, Penguin Books 1993 (430 pages). This is the classic "collapse" book, but written a decade ago before we used another 300 billion barrels of oil and now clearly recognize "peak oil."

The Essential Exponential for the Future of Our Planet by Albert Bartlett, Univ. of Nebraska, Lincoln, 2004 (291 pages). Professor Bartlett is our modern day Malthus but with a complete understanding of fossil energy. This book is a compilation of his best published papers and 1,560 lectures given in 49 states and countries over the last 30 years.

The Rapid Growth of Human Populations 1750–2000: Histories, Consequences, Issues Nation by Nation by William Stanton, Multiscience Publishing Comapny, UK, 2003 (229 pages). This amazing book is the most comprehensive compilation of world population numbers, history, and related issues. It is unimaginable that anyone could argue with the statistical case this author has developed, concluding with the inevitable crisis facing the world if population is not reduced. . . starting now (see additional comments in the Preface for the Third Edition.)

The Greening of Faith: God, the Environment and the Good Life edited by John Carrol,, Paul Brockelman and Mary Westfall, University Press of New England, 1997 (225 pages). Combined with his other books from the same publisher, Doctor Carrol clearly shows the conflict between stewardship of the planet and human tendancy to overpopulate and destroy resources.

Boiling Point: How Politicians, Big Oil and Coal, Journalists, and Activists Have Fueled the Climate Crisis—and What We Can Do to Avert Disaster by Ross Gelbspan, Perseus Books, 2004 (254 pages). This is the second book on global warming and climate change written by the foremost long time reporter on this subject. Gelbspan's first book, *The Heat is On* guided us into the long-term implications of fossil fuel combustion. His follow-up book is extremely disturbing as it makes the case that we are at or very near the tipping point for global disaster, with or without peak oil.

In my book (first edition), *The End of Fossil Energy and a Plan for Sustainability*, I purposely played down the problem of global warming because of its variability and camouflage by local and seasonal effects. After reading *Boiling Point* and considering our inevitable shift to coal, the last available and dirtiest of the fossil fuels, I am now a believer that if peak oil and natural gas don't get us first, we are in the crosshairs of climate change (see "triple-crisis" in the Preface to the third edition.) The poor Inuits are struggling for survival as the average arctic-winter temperature has increased almost 7°f in the past 50 years. See **www.earthjustice.org/inuit** or visit **www.giss.nasa.gov** (check "Global Surface Temperatures 2004") for the visual testimony of global temperature rise and anomalies since 1951.

Oil, Jihad and Destiny: Will Declining Oil Production Plunge Our Planet into Depression? by Ronald Cooke, self-published, 2004 (209 pages). This book is another excellent example of new input, creative thinking, and investigative journalism by an individual so concerned with our total world situation that he felt compelled to speak out. As an economist, Cooke's input breaks ranks with the usual "flat earth economist" (Colin Campbell's phrase) thinking and recognizes peak oil as a finite resource underlying most of society's present and future ills. He ventures bravely into the complex world of geopolitics, an area I purposely try to avoid because peak oil is truly apolitical. See **www.wizwire.com** for further information.

Food, Energy, and Society by David and Marcia Pimentel, revised edition, University Press of Colorado, 1996 (363 pages including 48 pages of references). This is the most comprehensive text that focuses on the inextricable relationship between energy, food, and population growth. I will reference this book later in this chapter while updating the ongoing controversy and the illusion of biofuels being our salvation in a post-fossil-fuel age.

Twighlight In The Desert: The Coming Saudi Oil Shock and the World Economy by Matthew R. Simmons, John Wiley and Sons, 2005 (432 pages). This book maybe the most profound exposé about peak-oil and fossil-fuel depletion. It is the culmination of Simmons' extensive study of 200 technical papers, which taken together, reveal the real truth about the precarious nature of Saudi oil reserves. Phrases like "fuzzy logic"; "giants at the tipping point"; "voodoo science"; heterogeneous carbonate reservoirs"; "coming up empty in new exploration"; and "will have to be rationed"

(p. 347) set the tone of this study. A most revealing Appendix C delineates Senate hearings in 1974 and 1979 "the smoking gun," which dealt directly with questionable Saudi oil capabilities. Unfortunately, the public never heard about those "alarm bells" from 30 years ago that Saudi Abrabia may have "reached its probable peak output and would soon need to rest its fields" (p. 384).

Energy for Survival: The Alternative to Extinction by Wilson Clark, Anchor Press, 1974 (652 pages). Another out-of-print, magnum opus, which was 30 years too soon. When the first oil crisis occured in the 70's, there was a great flurry of concern as typified by President Carter's attempts to do something. Unfortunately, new discoveries like Prudhoe Bay (Alaska) and the North Sea, plus continued surplus in the Middle East and Russia were more than adequate to supply all we could use, and the party continued for another thirty years. What a different world it would be if only we had heeded Clark's (and Carter's) words.

GeoDestinies: The Inevitable Control of Earth Resources Over Nations and Individuals by Walter Youngquist, National Book Company, 1997 second printing (500 pages). After being scarce for sometime, this keystone book is again available to lead us into the dangerous post-carbon world. It deals equally with the relationship between all mineral resources, the environment, health, water politics, conservation, and the "sustainable" society. This book also explains how severance taxes favor some parts of the U.S. like Texas, Alaska, and Louisiana with resultant constituancy and political repercussions. Also described are many short-term historic examples of "Boom and Bust" resource devastation from Montana copper mines to defunct iron-ore ranges in northeastern Minnesota. Younquist explains why even hydropower and geothermal are not truly renewable-energy sources because of long-term dam silting and the inevitable cooling of earth-thermal sources. These are more reasons for us to focus on immediate conservation and the quickest path to solar and wind.

Over a Barrel: A Simple Guide to the Oil Shortage by Tom Mast, Greenleaf Book Group, 2005 (111 pages). This is a concise book, self-published by a retired engineer, who was compelled by the looming-energy crisis to speak out and reach the public. (See **www.overabarrelbook.com** for ordering information.)

Energy Power Shift: Benefiting from Today's Technologies by Barry Hanson, Dakota Scientific Press, 2004 (204 pages), a very comprehensive-quantitative study of all the renewable clean-energy possible alternatives, which might lead to jobs and survival in the post-fossil-fuel age. Most of the concepts are legitimate. I disagree, however, with his wildly optimistic proposals for fueling our future with "billions of tons of free energy" from waste and biomass or "restaurant grade." He also devotes many pages about the promise and wonders of fuel cells. Where does the fuel come from? Natural gas? Windmills making hydrogen at 50% efficiency? I wish I could share his optimism.

The Hype About Hydrogen: Fact and Fiction in the Race to Save the Climate by Joseph Romm, Island Press, 2004 (237 pages). Finally, here is a complete book by an industry expert who explains why hydrogen is not the replacement for fossil fuels. Where I disagree with Romm is his minimal appreciation for peak oil and the urgency of fossil-fuel depletion. Instead he regards climate change as our most urgent problem.

The New Great Game: Blood and Oil in Central Asia by Lutz Kleveman, Atlantic Monthly Press, 2003 (287 pages). This is a new book by a war-zone reporter which clearly explains the pivotal role the entire Caspian region will play in the imminent world-oil end game. Kleveman personally visited British Petroleum headquarters in Baku when the decision was made to build the 3.2 billion dollar, 1090 mile, 42 inch pipeline from the Caspian Sea to Ceyhan on the Mediterranian Sea. The pipeline was just finished in May 2005, and not everyone in the Middle East and Asia is happy about it. It will be surprising if the world's oil policemen will be able to protect the flow of a million barrels a day (about one percent of world consumption) through mountainous and hostile country.

Plan B: Rescuing a Planet Under Stress and a Civilization in Trouble by Lester Brown, W.W. Norton, 2003 (283 pages). Brown is a renown author, environmentalist, and founder of the Earth Policy Institute. This is about his fifteenth book on world environmentalism. I agree that climate change and population stabilization are two of the greatest challenges facing civilization. Unfortunately, Brown does not address the critical importance of fossil-fuel depletion except indirectly as it relates to carbon emissions. Brown goes on to perpetuate the mythical panacea of hydrogen electrolysis using the power from wind, solar, or geo-thermal. In my opinion he also misleads his readers by taking a dead-end route and arguing that water and land productivity must be increased (Plan B) without explaining how this can be done with declining fossil fuels.

The Last Hours of Ancient Sunlight: The Fate of the World and What We Can Do Before It's Too Late by Thom Hartmann, Three Rivers Press, 2004 (379 pages). No fossil energy-ecology-population library is complete without this beautiful book by a well-respected author and advocate. The "voice of the earth" is trying to speak to us and from the back cover; "The more of us who heed Thom Hartmann's passionate message, the more likely it is that our highest hopes, rather than our darkest fears will come to pass." (John Robbins; founder of Earth Save.)

Strangely Like War: The Global Assult on Forests by Derrick Jensen and George Draffen, Chelsea Green Publishing, 2003 (185 pages). A comprehensive and disturbing book which describes how mankind has devastated the forests from ancient middle east and Greece through Europe in the middle ages to comtemporary rain-forests. The demise of fossil-fuels will only shift energy dependence back to wood, which is invariably de-pleted faster than it regrows.

Biodiesel: A Growing New Energy Economy by Greg Pahl, Chelsea Green, 2005 (279 pages). Another new book (somewhat like *Energy Power Shift*) touting biofuels as the answer to our prayers. This concept is at odds with the quantitative critical analysis (see Pimental above) of substituting fuel for food. There is a certain legitimacy of using soybean oil for transportation or heating fuel, but there are many negative considerations that need to be understood. This is such an important controversy (like hydrogen) that I address it again later in this chapter.

Energy: The Solar Hydrogen Alternative by John Bockris, Wiley and Sons, 1977 (364 pages). I was finally able to obtain a battered copy of this out-of-print book first printed in Australia. This is a most comprehensive technical textbook, which lays out 30 years ago the good, bad, and ugly about hydrogen. More importantly, it is a most accurate, prophetic analysis of the future for all the world's mineral and energy resources. Because of its thirty-year head start, it helps prove the invalidity of hydrogen. In other words, after thirty years, the same basic problems with hydrogen remain unsolved.

The Solar Fraud: Why Solar Energy Won't Run the World by Howard Hayden, Vales Lake Publishing, 2004 (281 pages). This book is a comprehensive summary which explains in clear quantitative language why non-fossil energy sources do not come remotely close to perpetuating our short industrial age. Unfortunately, Prof. Hayden does not include alternatives which might give hope and direciton to Americans just waking up to Peak Oil. Some day if we're both still alive, I'd be glad to give him a ride in my solar-powered MG.

Energy at the Crossroads: Global Perspectives and Uncertainties by Vaclav Smil, MIT Press, 2003 and paperback 2005 (427 pages, including 1,000 references). This is a most comprehensive book from an author capping a forty-year career of "systematic energy studies" beginning in Prague's Carolinum University. Smil has written eighteen books on energy, environment, and agriculture. His book, *Energies,* is referenced in the first printing of my book. He is obviously a world leader on all related subjects and one of my most respected mentors with serious concerns about CO_2 and climate change.

I agree with him on most points especially with his negative opinions of biofuels as our salvation, hydrogen with all its drawbacks, and his disdain for "Lovinsian Hypercars" and "Lovinsian exaggerations" (p. 322 referring to Amory Lovins of the Rocky Mountain Institute).

Where I disagree with Smil is his downplay and lack of concern about peak oil (like Romm and Huber/Mills). He references

Deffeyes' and Campbell's teachings of Hubbert's Peak but then questions their hypotheses with complete chapters, "Against Forecasting" (Chapter 3) and "Fossil Fuel Futures" (Chapter 4). I don't understand his reasoning except possibly he desires to give equal weight to referenced, optimistic forecasts by the USGS and P.R. Odell, *The Future of Oil* (a book I have not read). These are typical optimistic forecasts predicting two trillion barrels remaining, instead of one billion like Campbell, which shifts peak oil towards 2025 and serious decline by 2050. I wish these optimists were accurate, but with so many more experts disagreeing, we'd better plan for the worst as re-emphasized in Chapter 3. Other issues I have with Smil are:

1. His complete use of the metric system, which is meaningless to the American public. Terms like exajoules (EJ) and hectares (ha) belong in a college physics class.

2. His continuous confusion between efficiency and consumption. Like Huber/Mills in *The Bottomless Well*, Smil infers that steady increases in efficiency will save us, but then he reneges and admits that increased efficiency may also increase consumption.

As an engineer, I use efficiency by definition as the ratio between output energy and input energy in a machine or conversion process. In other words, a larger car (or power plant) may be **more efficient** in converting fossil fuel into work or electricity but at the same time **consume more** fuel because of its size. What we desperately need in the future, and will get whether we like it or not because of depletion, is **decreased fuel consumption**. In some cases this could be because of continued **increases in process efficiency**. But, more likely future **reductions in consumption** will happen because of reductions in usage (smaller cars, fewer trips). Most of the efficiency improvements have already happened. **Decreased consumption** is the precise basis of my Five Percent Plan.

I belabor this issue of semantics because it is a primary source of confusion between experts, authors, media, politicians, and the public. They hear a mixed message, thus squandering our focus and precious time left, if any, to plan for the coming fossil energy crisis.

The Bottomless Well: The Twilight of Fuel, The Virtue of Waste, and Why We Will Never Run Out of Energy by Peter Huber and Mark Mills, Perseus Book Group, 2005 (214 pages). This strange book was just published and promises everything is fine. The more we use the more we'll have. Growth, consumption, and Utopia will go on forever in the veil of "infinite logic." Unfortunately, the authors' curves of parallel fossil energy, GDP, and employment end at their year 2000 level, and then conveniently extrapolate onwards and upwards just like the EIA forecasts or any similar pre-peak curve of oil and energy production in lock-step with consumption and growth. I would recommend this book to a high-school physics class with the challenge to look for obfuscation and distortion in science, math, and logic. Irrelevant diversions into DNA science, computer chips or laser beams will not answer our coming energy and power problems.

This concludes the book review portion of Chapter 2. We will now pick up the original Chapter 2 with an update of non-fossil energy sources.

ENERGY SOURCES
OTHER THAN FOSSIL FUELS

SOLAR

Incoming solar energy is still our best hope for an acceptable industrialized future. It is clean and ubiquitous but dilute even when the sun is shining. Solar energy arrives at the earth's surface as radiation from nuclear fusion in the sun. The maximum power intensity that this energy flows in perpendicular to the earth's surface is a rate of about one kilowatt per square meter (10.9 square feet) of surface area. Because of efficiency losses, the actual availability of this power is significantly less. With optimal conditions, photovoltaic cells will convert only about one-tenth of this solar power to electricity or 100 watts per square meter or about 10 watts per square foot. Non-electrical but active- or passive-solar heating for hot water or residential heating can be up to twice this amount (200 watts/m^2) under ideal conditions. In real life, the total usable solar power varies greatly with latitude and hours of unimpeded sunlight.

Typical useful solar energy for electricity or heating is between 2 to 8 hours per day. This is called insolation or equivalent hours of full direct sunlight. The actual, useable energy would therefore vary from 200 watt hours per day to 800 watt hours per day for each square meter of photovoltaic solar panel.

One very important consideration as we move to a civilization fueled by incoming solar energy is the considerable energy required to melt and process the silicon into photovoltaic cells. Initially, this input energy was more than the cell would produce in its lifetime. With steady improvements, this pay-back period can now be lower than one year, which means that it takes less than the first year of electrical output just to return the energy to make the solar cells. Of course, this manufacturing energy now comes conveniently from fossil fuels, but eventually as we run out of fossil fuels the photovoltaic manufacturing process will have to provide its own energy for production.

By 2005, there has been an explosion of interest in solar energy. The problem is that all today's solar energy is only one-sixteenth of one percent of our total energy usage. There is no hope of ever reaching present energy levels especially in the short time we have. Suddenly, because of awakening interest, new panels are becoming scarce and more expensive. New thin film technology from companies like Konarka, Nanosolar, and Nanosys show promise but are a long way from quantity production and may not have the 50-year life or efficiency we expect with silicon cells.

On the bright side, countries like Japan and some states like California have committed to significant solar tax incentives in the next decade. Solar is a tiny step to renewable energy even if electricity will not run today's cars or fly our airplanes. In fact, the total U.S. solar energy produced in 2003 including non-photovoltaic thermal solar is about equal to 400 million gallons of gasoline used in one day (0.01 EBBO).

For detailed information on solar energy, visit the website for the American Solar Energy Society, **www.ases.org** and the bi-monthly periodical **www.solartoday.org**. The Jan/Feb 2005 issue features comprehensive articles like "Japan Takes the Lead … and Leaves the United States in the Dust." This article shows

Japanese solar panel production approaching 700 megawatts in 2003 compared to 100 megawatts for the U.S.

On the contrary, in this same issue, an article by Donald W. Aitken Ph.D. is titled "The Renewable Energy Transition: Can It Really Happen?" The conclusion is yes but only with completely unrealistic (my opinion) increases of solar, wind, and biomass. Dr. Aitken's projected world production increases to 30 EBBO for solar and 20 EBBO for wind are impossible when compared to the 2003 U.S. production of 0.010 EBBO for solar and 0.018 EBBO for wind. In addition, Aitken forecasts a peak of all world fossil energy in 2030 and a 2050 level similar to 2003 of about 65 EBBO. This is a far different scenario than projected by ASPO (The Association for the Study of Peak Oil and Gas) **www.asponews.org** and more typical of optimists like the EIA and USGS.

Aitken's prophecy of biomass energy increasing to about 15 EBBO worldwide by 2050 is totally at odds with the inevitable demand for food and dwindling supplies of oil and natural gas to support the food and/or biomass production.

I dwell on this particular article because it is a good example of the optimistic message projected by the NREL (National Renewable Energy Laboratory, wing of the DOE). This is the same government agency that plans to use wind energy to electrolyze hydrogen for our transportation fuel. There are more than enough qualified dissenters to raise serious questions about these proposals. Meanwhile, the public and our elected leaders hear these optimistic proposals, and we lose precious time while the world continues year after year using one billion barrels of oil every eleven days and moves inexorably ever closer to the demise of industrialized civilization. For further reading see Hayden, *The Solar Fraud* reviewed on page 32.

WIND

Another form of solar energy is from air movement caused by solar heating and cooling of the atmosphere. The available energy from wind impinging on a rotating blade attached to a generator will vary greatly depending on wind speed and blade design. A typical value (similar to photovoltaic power) is 100 watts per square meter of swept blade area. Again, like all solar energy, a considerable amount of fossil fuel is needed to make the blades, generators, towers, etc. In a fossil-fuel-free civilization, all of this manufacturing energy will have to come from renewable sources.

In 2003, the total annual wind energy in the U.S. (0.018 EBBO) was equivalent to less than two day's gasoline consumption in our automobile engines. I make this point to explain the impossibility of using wind energy (via hydrogen at 50 percent efficiency) as a transportation fuel. As oil and natural gas supplies dwindle, we will need all of the wind energy we can generate to help supply our electrical needs.

The Jan/Feb 2005 issue of *E Magazine* (**www.emagazine.com**) is devoted almost entirely to wind, "The World's Fastest-growing Renewable Energy Source is Coming of Age." Like Japan with solar, Germany leads the world with 14,600 megawatts of wind power installed by the end of 2003. Denmark has the highest proportion of capacity with 20% of electricity already generated by wind. For comparison, the U.S. is projected by the AWEA (American Wind Energy Association) to have 10,000 megawatts in production by 2009. If all of these projected U.S. windmills produced full power 25% of the time, they would provide the same electricity as 0.031 billion barrels of oil burned in a power plant at 35% efficiency. This is still only about 0.2% of our 2003 total energy consumption of 15.7 EBBO and about equal to three days of present gasoline consumption.

I continue to dwell on the magnitude and quantitative comparison of renewable versus fossil-fuel energy because it sounds like solar and wind are huge, making great progress, and will save us; whereas, in reality, they are miniscule and puny when compared with the fossil energies and need considerable input energy for their manufacture.

The journal of the optimum population trust in UK as edited by Andrew Ferguson gives the most quantitive, cogent arguments why wind and solar energy can, at best, supply only a small sporadic percentage of our future needs unless we reduce population from seven to two billion people (see website on page 61).

BIOMASS

A third form of solar energy is photosynthesis resulting in biomass (living matter). The ubiquitous green in nature indicates the conversion by incoming solar radiation of CO_2 into more concentrated, carbon molecules that can be used as fuel. As pointed out earlier, biomass in various forms has been a source of civilization's energy requirements for thousands of years. The problems with overdependence on biomass are many. Obviously, it takes proper soils, water, and time to grow the organic fuel. Wood, in Europe, was seriously depleted in the Middle Ages for fuel and construction. We've had a reprieve in some areas with the advent of the fossil-fuel age, but in most countries, the availability of fossil fuels has encouraged serious continued deforestation from the rainforests of the tropics to Third World temperate zones. One example of an all-out assault on the earth's resources is the use of rapidly dwindling fossil fuels to cut, chip, transport, and process wood to generate additional electricity referred to as renewable "green" energy. Wood is not truly a renewable energy if we do not follow a very careful plan to conserve our forests and if fossil-fuel energy is required for harvest and transport. In Chapter 7, I describe a microcosm of the depletion of biomass as it occurred on Easter Island. Unfortunately, in the last two years, biomass in its many forms continues to be offered as a the answer for our coming energy crisis, but the basic problems remain.

The use of input fossil energy (or some fraction of the output bioenergy) to fertilize, till, irrigate, transport and process the biofuel must be deducted from the gross-energy output. Some think tanks like ILSR (Institute for Local Self Reliance, "How Much Energy Does It Take to Make a Gallon of Ethanol," 1995, Lorenz and Morris) argue that with modern methods, and if it's done just right, the energy output of ethanol and co-products is greater than the input. On the contrary, the recognized expert on this subject, Cornell University professor David Pimentel (see "Ethanol Fuels: Energy Balance, Biomass and Environmental Impacts are Negative," *Natural Resources Research*, volume 12, number 2, June 2003 and his book, *Food, Energy, and Society*, reviewed earlier in this chapter) has been teaching and writing for years that growing biomass for transportation fuel is absolutely faulty thinking. In addition to the negative return on energy, competition for food and topsoil degradation also has to be considered. In some cases, large, government-subsidized corporations move in and displace the small farmer. This has happened in semi-tropical Brazil where ethanol from sugar cane in place of devasted forests is offered as proof of energy from biomass. Closer examination shows the socioeconomic ills which occur when a mono-biomass culture trumps diversified indigenous food production.

Still the controversy goes on. It is technically true that biofuels, including biodiesel, could power our cars and even airplanes or heat our homes, but it should only be in dire emergencies and at the serious expense of food production just when inputs from fossil energy and nitrogen fertilizer from natural gas wind down. All this will happen as we have depleted our topsoil and are trying to feed a fossil-fueled, six-fold increase in population.

We cannot deny the basic science and math. Incoming solar energy is very dilute. It takes considerable arable-land area to provide meaningful quantities of biomass energy. For example, one acre of soybeans, even at a 4:1 positive-energy yield would provide up to 75 gallons of biodiesel fuel (Pimental, *Food, Energy, and Society* p. 126). If we were to use this as a substitute for **only** 5% of our annual 1 EBBO diesel fuel **or (not and)** 1 EBBO heating oil consumption about 28 million acres of cropland would be required. The use of rapeseed (canola oil) might double the yield up to 150 gallons per acre. There is some promise of algae providing algal-biodiesel at considerably higher yield per area of water surface. The National Energy Research Lab (NERL) has run experiments indicating up to 2,000 gallons of biodiesel per year from 1,000 squared meter ponds. Maybe this is our energy future but much work remains to be done.

There is no question that the greatest crisis caused by the peak and decline of fossil fuels will be the concurrent decline of food. The world's population increased last century in lock step with fossil energy and food production. In the future, every unit of bioenergy will be required to feed an overpopulated world.

   World population, now in excess of six billion, will no
longer be sustainable without the interwoven food/fossil
fuel fabric, which allowed it to happen. The poorer
countries are clearly at the edge. The industrialized
countries can only follow. Food just cannot be grown,
processed, and transported at today's level without
fossil energy.

HYDROELECTRIC

The sun's energy warms the air which evaporates water in the
oceans. Warm air rises and carries the water vapor to higher, colder
elevations where the moisture is squeezed out and returns to the
earth in familiar forms of precipitation. During the gravitational
flow back to lower surfaces, water accumulates and can provide
significant energy to generators conveniently placed at dams, which
control flow quantity and direction. Hydroelectric power is an excellent source of clean energy although fish and locally displaced
people may not agree. Also, true sustainability is questionable because of silting. However, it does not rely on finite-fuel sources, except as required to build and maintain the dams and generators and
it is a fact that hydropower can provide considerable electricity for a
modern civilization—but only where the water flow is adequate and
most of the accessible sites have already been utilized. Today, about
7% of the total energy in the world comes from "hydro" and even
that amount is not entirely dependable because of the weather. Between 1996 and 2001, the U.S. experienced a 40% decline in hydroelectric output because of drought conditions (probably caused
by global warming). Another indication of the limitations of hydroelectric power is that it has decreased from providing about 40% of
U.S. total electric generating capacity in 1930 to 3% in 2003. Demand sharply increased while hydropower remained flat. In fact, ac-
cording to EIA Table 1.3 the 2003 production of 0.44 EBBO is almost exactly the same as 0.45 EBBO in 1973.

NUCLEAR

Just after the beginning of the 20th century, Albert Einstein proposed that mass and energy are equivalent. This is shown in the

familiar equation: $E = MC^2$ where E is energy, M is mass, and C, the third term, is the speed of light (186,000 miles/sec). Only certain isotopes (rare higher molecular weight atoms) of scarce materials like uranium can be used to fuel a nuclear bomb or power plant. The only proven process is called fission, which releases great amounts of deadly radiation as a byproduct. Contaminated materials and spent fuels must be carefully secured and stored. As recently as 1970, nuclear power was expected to be the answer for an energy-addicted civilization. Subsequent disasters at Three Mile Island in Pennsylvania (March, 1979) and Chernobyl in the Ukraine (April, 1986) completely turned public and considerable scientific opinion against this direction. Today, some countries like France continue to carefully produce a large share of their electrical power from the fission of uranium. In the U.S., about 20% of our electrical power and 8% of our total power comes from nuclear energy.

There is talk in the present Bush administration of increasing our reliance on nuclear energy, but the problems of waste disposal have not been solved, and even if all the environmental hazards could be resolved, the fuel sources are finite and were heavily mined during the Cold War. The World Nuclear Association[2] paints a rosy picture about the future of fission and uranium sources. We must continue to keep in mind, however, that the mining of uranium is very environmentally destructive and requires huge amounts of fossil fuel to extract, move, and process. Eventually, nuclear energy will also have to provide its own up-front energy to be truly sustainable. How will the earthmoving and processing equipment be fueled?

Nuclear fusion, like the sun, is the Holy Grail of energy and is being offered as the hope for an ultra-high-tech civilization. To date, after many years and billions of dollars of research, we don't seem to be any closer to a practical power plant. Maybe, if we could keep our industrialized civilization going for another 150 years, we could find a path to a fusion-powered world (other than the old reliable sun). By 2003 the annual U.S. nuclear power production of electricity has changed little from 1.28 EBBO to 1.27 EBBO. Nevertheless, the hope of fusion energy does not die. There are experts in the U.K. arguing strongly that we could

41

revive nuclear power if we had the time and focus (contact Brendan McNamara: brendan@leabrook.co.uk).

OTHER ENERGY SOURCES

There are several other energy sources that are either in use, under development, or have some relevance. These are geothermal, which utilizes the heat under the surface of the earth, and tidal, which uses the gravitational pull of the moon to harness tidal water flow.

There have even been proposals for harnessing the difference in seawater temperature between the ocean surface and very low depths and converting this into electricity. There is a surge of interest in harvesting wave power. All of these infer highly localized sites or speculative technology and to date have little effect on our total energy requirements. For instance, the 665-megawatt "geysers" plant in California, the largest geothermal plant in the U.S., has the capacity to produce electricity equal to about one-third of the 0.01 EBBO total solar production of 2003. Again, if we could buy some time with drastic conservation of fossil energy, we might find the time to optimize and further develop these minor but legitimate alternatives.

There is also a fringe world devoted to possible energy sources, which include the bizarre to the real to the radical. I'm speaking of mysterious concepts like electromagnetic motors, gas (methyl) hydrates, cold fusion, and deep abiotic (non-fossil) petroleum origins. This is not the place to totally investigate or debunk each of these theories. I wish at least one could save us from our post-carbon fate, but there does not appear to be a sliver of hope for any of them in the time we have left. Some are so technically complicated to be incomprehensible or infer perpetual motion or "something for nothing."

For the reader so inclined to pursue such concepts, I offer several starting points:

- **Gas Hydrates** (sometimes referred to as gas clathrates or methane hydrates)—These mysterious crystalline solids exist in great quantities under deep layers of permafrost or in oceanic locations at depths between 2,000 and 8,000 feet where pressure is high and the water is within a few degrees of freezing. There has been intermittent methane production from one site in Siberia since 1969, typical energy production of a mid-sized Texas gas well.

It may be conceivable that there are vast reserves of energy locked up in hydrates, but as my number one information source on this subject (Kenneth Deffeyes' book, *Beyond Oil*, pages 71–76 referenced earlier in this chapter) states, "It won't be easy; the major oil companies have known about the gas hydrates since 1970, and so far they haven't announced a promising extraction strategy." Another good source for up-to-date activity regarding methane hydrates as well as all energy issues is **www.energybulletin.net**.

- *Excess Fusion, Why Cold Fusion Research Prevailed* by Charles Beaudette, Oak Grow Press, South Bristol, Maine. I have no comment. I have not read this book. Still, the dream lives on.

- *Abiotic Oil: Science or Politics* by Ugo Bardi (**www.aspoitalia.net**). The subject of abiotic oil has been around for some time. The theory was initiated by Russian and Ukrainian scientists in the middle of the 20th century and has resurfaced from time to time with no credibility by oil industry experts. It definitely is not helping in oil-depleted U.S. and may be just terminal seepage from old sources. (See **www.gasresources.net** for unlimited controversy on this subject. Enjoy.)

- **Electromagnetic Radiation, U.S. Patent 5,590,031** (**www.uspto.gov**)—awarded in 1996 to Mead and Nachamkin for a system for converting electromagnetic radiation energy to electrical energy. As Tom Beardon comments, "It's legitimate, but nothing will ever power anything of substance, not even a watch. But the principle is there!" Tom Beardon's website for many exotic energy sources is **www.cheniere.org**.

- **Motionless Electromagnetic Generator, U.S. Patent 6,362,718**—was awarded in 2002 to Patrick et al. for a motionless electromagnetic generator (MEG). From the patent abstract, this is a rotary machine that apparently produces power from the residual flux in the permanent magnets. I cannot begin to understand the patent itself. Read in the following Russian magazine about a Japanese manufacturer that is making small fans using this theory. (For further information see **www.cheniere.org** as well as the next item.)

- *New Energy Technologies*—A most enlightening English language magazine published in Russia for an annual subscription of $49.99. All back issues can be obtained as PDF files on a $29.00 CD (see **www.faraday.ru** for ordering information). You can read about everything from thermolevitation to multi-rotor homopolar devices as well as more understandable articles about basic world-energy problems (for a personal contact, try the editor, Alexander Frolov at his email address: office@farady.ru).

HYDROGEN

The proposal of hydrogen[3,4,5,6,7,8] as the fuel of the future and solution to our coming energy crisis is misleading and dangerous because it gives a false sense of optimism. Liquid hydrogen has a very high-thermal energy content of 61,000 BTU per pound compared to about 18,000 BTU per pound for common liquid fuels, but the problems with hydrogen are many:

- It is a very flammable light gas and requires extremely high pressures (about 5,000 pounds per square inch) or very cold temperatures (minus 400°F) to liquefy it for storage or shipment. Compare these hurdles to today's procedures and safeguards necessary for bottled propane at about 200 psi.
- It is a very common element but exists chemically bound with oxygen as water or with carbon in fossil fuels. It also appears in many other chemical compounds but always combined tightly with other elements and requiring considerable energy for separation.
- Of all hydrogen used today, about 97% comes from fossil fuels (primarily natural gas) and 3% from electrolysis of water. What sense does it make to use our finite-fuel sources just to make a different form of energy? In the case of electrolysis, it takes more energy to separate the hydrogen than is returned when the hydrogen recombines with oxygen resulting in about 50 percent efficiency. When hydrogen is made from

the high temperature stripping of natural gas, considerable provision must also be made for disposal of the resulting (sequestered) CO_2.

For practical purposes, hydrogen is a misleading and delusive issue in our coming fossil-fuel crisis. It is difficult to believe that it can be proposed as the solution. Our decision makers hear optimistic predictions about hydrogen and divert attention from the real problem, the finite reserves and depletion of fossil fuels.

The best hydrogen can do is to provide a very exotic, expensive storage medium (like a super battery). The stored-energy density by weight of liquid hydrogen at 5,000 psi is three times higher than that for liquid fossil fuels, but this is offset by the weight of the high-pressure container and four times larger volume required. Hydrogen-fueled cars and airplanes have been prototyped for many years, but cost and complexity have limited these efforts to technical oddities with huge expensive fuel tanks. In April 1988 the Soviet hydrogen-fueled TU-154 jetliner used a large portion of the fuselage for the hydrogen-storage tank to power just one of three engines. Supporters of hydrogen refer to this experiment to argue that hydrogen can fuel commercial and military aviation without clarifying the compromises that have to be made.

It is true that liquid hydrogen could be a carrier of concentrated energy similar to fossil fuels. In a totally fossil-fuel-free civilization and with electricity available only from renewable sources, liquid hydrogen may be the only concentrated energy other than biofuels when and where requirements cannot be served by batteries or a long extension cord. Examples might be an agricultural tractor or construction machine requiring huge amounts of power to build a dam or extract uranium. A typical 25 kw (33 Hp) tractor, similar to a small 2-plow farm tractor, could be fueled at the farm with hydrogen produced from photovoltaic electricity. At one kw peak power from each ten square meters of grid, a 250 square meters (about 50 feet by 50 feet) array would provide one hour of tractor power at full power for every hour the grid is exposed to direct sunlight. This is equivalent to two gallons of diesel fuel or gasoline burned at 25% efficiency. Such a photovoltaic array alone would cost $100,000 at today's prices.

This is in addition to the hydrogen-handling equipment and fuel-cell, electric-drive complexity if used instead of a conventional piston engine. Biofueled or battery-powered tractors and even draft animals seem far more realistic. By the same thinking, a typical 100 kw bulldozer would require four times greater investment in photovoltaics and hydrogen handling equipment, but it will be the only way to do heavy earth moving other than with biofuels. We will still need petroleum-based lubricants, hydraulics, and plastics or rely on substitutes made from refined biomaterials.

Yet, millions of dollars are being awarded to fund hydrogen research. A large part of this investment would be better spent to develop energy-independent personal residences and electric vehicles for the future as well as to explain, subsidize, and encourage the conservation of the fuel we have left. A comprehensive analysis of hydrogen problems can be found in "Why Hydrogen is No Solution, Much Ado About Nothing" and "Why Hydrogen is No Solution—Scientific Answers to Marketing Hype, Deception, and Wishful Thinking." Both can downloaded from *From The Wilderness* newsletters (**www.fromthewilderness.com**).[9]

By 2005 little has changed concerning the hydrogen hype. Hundreds of millions are still being spent to prove it will work. A very recent example can be found in *Scientific American*, June 2005, titled "Solid (State) Progress." This article describes the development of "cryoadsorption," which relies on getting hydrogen to adhere to other materials at easier to attain temperatures of -196°C and pressures of 1,000 psi. Many sceptics are still trying to explain why a hydrogen-fueled economy cannot happen while precious time is wasted. The book, *The Hype About Hydrogen*, is reviewed earlier in this chapter.

Two excellent recent summaries of why the challenges facing civilization cannot be solved with hydrogen can be found in two articles by Dale Allen Pfeiffer on **www.321energy.com**. See The Myth of the Hydrogen Economy (Jan. 5th 2006) and The collapse of Complex Systems (Jan. 14th 2006).

Chapter 3

Where Are We Now?
A Situation Analysis–2003

Including World and U.S. Energy Consumption and Internet References

Our short fossil-fuel era is defined by previously stored energy. For hundreds of millions of years organisms lived and died and were subsequently compressed with heat into a solid, liquid, or gas, which were buried into layers of the earth's crust. This concentration of hydrocarbons resulted in stored energy, the basis for our modern Industrial Age. Fossil fuels are not abundant throughout the world. An excellent, in depth analysis of how oil is formed, where to look for it, and how much is left can be found in the book, *Hubbert's Peak*,[1] written by Kenneth Deffeyes, a life-long petroleum geologist and professor emeritus of Princeton University. Another new book, *The Essence of Oil & Gas Depletion*,[2] has just been published by Colin Campbell, the recognized world leader on this subject. This comprehensive book describes in detail the oil-production history and projected reserves for every important country (see Appendix 5). It also includes timely updates of many aspects of the title subject. Quoting from the back cover of Campbell's book:

Oil and gas are finite fossil fuels from the geological past that are inevitably subject to depletion. Eventually we must run out, but what matters more is the inevitable peak of production when growth gives way to decline. The wider implications of this historic discontinuity are colossal.

In recent years, experts have learned how to more accurately quantify how much fossil fuel we have used and how much is left. After starting with coal, which is the most accessible, we are now rapidly consuming the cleanest and most available oil and natural gas. All three forms can be used somewhat interchangeably. Heavier forms of oil and coal can be converted to gaseous or liquid forms, although with attendant energy input and pollution.

In order to combine information from many sources into a quantitative and more understandable format, I will use the following methodology:

1. Start with total world energy consumption in all forms—fossil fuels, nuclear, and renewable.
2. Convert all energy forms to equivalent barrels of oil. Because of the huge quantities involved, the common term used will be equivalent billion barrels of oil (EBBO). If electrical output is the quantitative value of energy such as from a nuclear plant, a hydroelectric dam, solar cells, or a wind farm, the electrical output in kilowatts will be divided by an efficiency factor of 35% (0.35). This equals the equivalent billion barrels of oil (EBBO) that would be required to produce the same amount of electricity in a conventional power plant.
3. Use Department of Energy/Energy Information Administration (DOE/EIA)[3] statistics as the primary reference for past and present energy production and consumption in the U.S. and throughout the world.
4. Refer to other sources as needed for verification and analysis of remaining reserves.

From this information, the big picture is clear. We have become hooked on plentiful fossil energy. "Addicted to oil" (per Pres. Bush) in his 2006 state of the union address.

Earlier authors have anticipated the problem,[4,5,6,7] but surplus Middle East oil has kept us in a state of energy euphoria. Now the picture is becoming clearer. We are reaching and going over the peak of available resources and looking at the downhill side. Jeremy Rifkin calls this, "sliding down Hubbert's Bell Curve," in Chapter 2 of his book, *The Hydrogen Economy*.[8] For our analysis, I will intentionally de-emphasize the remaining coal and non-conventional types of oil in the form of shale oil and tar sand. There are loose estimates of reserves of these forms of fossil fuels, equal to hundreds of billions of barrels of oil, but they require huge amounts of energy to process and cause considerable environmental problems both in extraction and refining. Global warming and pollution will be even more serious as we inevitably are forced to accelerate dependence on these less desirable forms of concentrated fossil energy. As we struggle to reach a worldwide sustainable-energy future, these last forms of fossil fuel will be our final chance to continue a modicum of industrialized civilization. If we don't count them in our pocket now, we will have more incentive to start immediately to make the necessary reductions in oil and natural gas usage.

We will need all remaining forms of fossil energy for agriculture, national defense, modern healthcare, plastics, lubricants, hydraulics, absolutely necessary transportation, etc. We need to play a mind game and pretend the dirtier fuels are not available until we absolutely need them.

The following finite-resource consumption curves show world oil consumption to date as of 2001. See Chapter 2 for updates. This visual display is from page 5 of *Hubbert's Peak*.[1] Also shown on the same coordinates is U.S. oil production to date (from pages 143 and 155 of the same book). It is interesting to note that U.S. production, which is well past the peak at about 1971, lies exactly on the expected bell-shaped probability curve that world production will soon follow (see Appendix 5 for "A Numeric Summary of World Past Production and Remaining Reserves").

World and U.S. Oil Consumption Through 2001

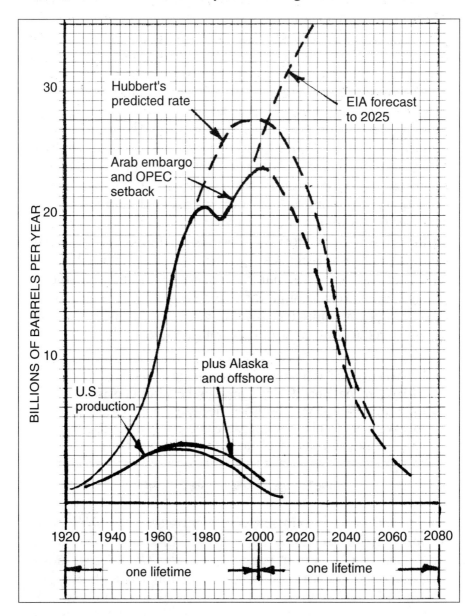

The dotted line continuing upward is the EIA forecast to the year 2025. This gross disparity between the petroleum experts and optimistic governmental projections is extremely important for our understanding of fossil energy and will be discussed further throughout this book.

The original, total U.S. endowment of oil was 220 billion barrels of which 200 billion barrels have already been used. This leaves 20 billion barrels left unused in the U.S. At over 6 billion barrels consumed annually, we have only a three-year supply of our own oil left. The U.S. Geological Survey (USGS) estimates that an additional 7 billion barrels of recoverable oil may be in the Alaskan National Wildlife Refuge (ANWR), enough for one-year consumption in the U.S. at the present rate. It would be folly for us to use this now when it will be so precious for our children. They'll wonder what we were thinking of when we wasted energy as we do today.

Of the original 2,000 billion barrel total endowment of world oil, about 1,000 billion barrels of oil have been used. Most experts (including those in British Petroleum and Royal Dutch Shell) agree that there are about 1,000 billion barrels left. We are half full (or half empty), but most of our consumption has been in the last 25 years. At the 2002 world consumption rate of 28.5 EBBO annually we will be out of oil in 36 years (see Appendix 5). Also, there is considerable uncertainty as to how much of the remaining oil will be extractable. And, at what price?

By 2005, when this book has been updated, we will have used another 55 EBBO. By 2006, Deffeyes tells us (now as a historian) that "we passed the peak of oil production on Dec. 16, 2005... by 2025 we're going back to the stone age" (**www.princeton.edu/hubbert** Feb. 11, 2006).

At this point, I will focus on the huge disagreement over the quantity of remaining reserves. The oil experts like Deffeyes[1] and Campbell[2] are telling us loud and clear, "We're going over the peak." They can predict this well in advance by analyzing the ratio of new discoveries to consumption. This ratio is already in serious decline, by a ratio of 1 to 6, a warning of the end for any finite reserve. This obvious mathematical prediction is totally at odds with the international energy outlook to 2025 as reported by the EIA.[3]

The EIA predicts in their 25 year forecast, shown extended beyond the above consumption curves, that world total energy production (including oil) will increase 58 percent by the year 2025. Oil production will increase by 70 percent from 25 EBBO to 42 EBBO per year. For this to happen the world would use 805 EBBO of oil in the next 25 years. If production then continued at a constant level from 2025 to 2043, we would use an additional 756 EBBO for a total consumption of 1.56 trillion barrels in the next 40 years, all from a reserve of one trillion barrels.

Someone is dead wrong on a very important disagreement. On such an important issue, it would seem prudent to err on the conservative side.

What difference does it make anyway, whether we have 30 or 50 years of oil left. The crisis facing us is still the same.

Natural gas reserves are difficult to quantify especially in less friendly countries, but they appear to be similar to or little better than for petroleum. World reserves are presently estimated at 155 trillion cubic meters or an equivalent 884 EBBO. World natural gas consumption rate is 14 EBBO per year; therefore, we would appear to have a 63-year reserve to production ratio (884 EBBO divided by 14 EBBO equals 63 years). Sixty-three years may seem like a long time, but this time period will be considerably reduced as consumption is increased. This is inevitable as petroleum becomes more expensive and especially if natural gas is stripped for a hydrogen economy or used to process tar sand. The EIA estimates natural gas usage will double in the next 25 years. Canada now sends over half of its natural gas to the U.S. but production is in decline.

Regardless of which estimates are used, we are talking about depleting nearly all of the world's oil and natural gas in far less than the lifetime of a child born in 2003.

As we go through 2003, even the above forecasts may be considerably optimistic. U.S. natural gas production has declined to an annual rate of 524 billion cubic meters (3 EBBO). In this same period, wholesale prices have more than tripled from about $2.00 per thousand cubic feet to over $6.00. John Wood, the director of the EIA in Dallas, says that the higher prices will increase exploration and supply. Michael Zenker, a senior director at Cambridge Energy Research Associates, counters that our cornerstone of gas supplies along the Gulf Coast is petering out. "We're having a tough enough time keeping up with declines from existing fields."[9] As of January 2006 the price exceeded $10.00 before dropping back to the $8.00 range due to a record warm winter.

Like oil, there is some disagreement about natural gas reserves, but remember, this is the same natural gas proposed to fuel a future-hydrogen economy. We're having trouble now just satisfying electrical and heating demands. Shortfalls of natural gas are already disrupting U.S. economy because liquefied natural gas (LNG) can only be landed at four U.S. ports. Additional LNG production will not solve our problems in time.

At this point, I will return to an analysis of present (as of 2001) production of all energy sources. Since this is being written in 2003, all U.S. production of 15.5 EBBO in 2001 will be rounded upwards to 16 EBBO. Oil, with a 39% share of total energy, is the primary fossil-fuel source being depleted today. The second and third sources are coal and natural gas, each providing about the same 23% share. Together, the big three: oil, coal, and natural gas, provide 86% of the world's energy.[3] From the DOE statistics, we can find the contribution of each energy source in the world and the U.S. for 2001. See Chapter 2 for 2003 and 2004 updates but with little change.

	WORLD		USA	
Oil	39%	24.88 EBBO	39%	6.05 EBBO
Natural gas	23%	14.8	24%	3.69
Coal	24%	15.25	23%	3.55
Hydroelectric	6.7%	4.27	3%	0.50
				(0.62 in 1996)
Nuclear	6.6%	4.21	8%	1.20
All other forms	<1%	0.49	3%	0.55
Total in 2001	64 EBBO		15.4 EBBO (in 2001)	
			16.0 (est. in 2003)	

All other forms in the above table includes wood, waste, solar, wind, geothermal, and tidal. A close examination of the above numbers shows some inconsistencies depending on which EIA source is used. For purposes of this book, direct electrical-energy production from non-fossil-fuel sources (hydro, nuclear, solar, and wind) will be divided by a 0.35 (35%) efficiency factor to find comparable, equivalent billion barrels of oil (EBBO). As stated earlier, the 15.5 total consumption was 2001 was estimated to increase to 16.0 in 2003 as the first printing of this book goes to press. (It actuality reached 15.75 by 2003.)

Of the 13.29 EBBO of fossil fuels consumed in the U.S in 2001, approximately 5% (0.59 EBBO) is for non-fuel use (not including liquid petroleum gas) such as asphalt, petrochemicals, plastics, lubricants, etc. These represent some of the thousands of ubiquitous fossil fuel uses (including hydrogen production) that we take for granted in this short fossil-fuel age. They are typical of the products that will become very dear as we move into a post-industrial age. Another critical need for fossil fuels is for national defense. In 2001, almost 1% (0.123 EBBO) of fossil fuels was used by the Department of Defense (DOD) with 72% of that used for jet fuel.

In the set of curves shown next, I've combined all world and U.S. energy consumption from the earlier table into similar bell shaped curves. These show we are at or near the top of Hubbert's

World and U.S. Energy Consumption from All Sources

(from the table on page 50)

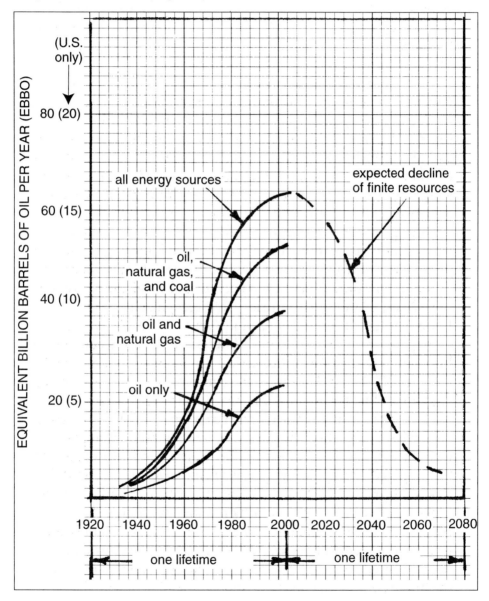

Peak as the whole world (led predominately by the U.S.) sits on the energy-age bubble. It is possible to put both U.S. and world consumption onto exactly the same curves just by changing the Y-coordinate by a factor of four. This is because world consumption is four times that of the U.S. for all major sources even though the U.S. has only one-twentieth the population.

If a picture is worth a thousand words, these consumption curves show our present-precarious position, how fast we got here, and the typical expected resource decline for all fossil-fuel sources. We are facing drastic changes in the time frame of our and our children's lives. There is no easy way out of this dilemma. Our only hope for an acceptable-future lifestyle is to address this situation immediately. Many authors and scientists have been calling for action, but little is happening.

A tragedy of our civilization is that, in the last 20 years in spite of cries for conservation, we have continued blindly in the wrong direction. Our leaders have acquiesced to the short-term gratification of constituents and even encouraged fossil-fuel consumption. Many European countries have recognized the challenge. Initiatives have been introduced in Germany and Italy to install solar panels on 10,000 homes. There is legislation pending in England to require 20% of electrical production be from wind by the year 2020. For decades European countries have had much higher gasoline taxation to discourage excessive consumption.

It is sad that Jimmy Carter tried to warn us of our coming energy crisis nearly three decades ago and couldn't accomplish his goals, which would have kept us out of our energy woes today. In his proposed energy policy speech of April 18, 1977, he said, "... this is the greatest challenge our country will face during our lifetimes," ... "we must not be selfish or timid if we hope to have a decent world for our children and grandchildren. We simply must balance our demand for energy with our rapidly shrinking resources. By acting now, we can control our future instead of letting the future control us. Ours is the most wasteful nation on the earth. The oil and natural gas we rely on for 75 percent of our energy are running out." Many 1985 goals were proposed including "solar energy on two and one-half million homes and a strategic oil petroleum reserve of one billion barrels, more than six month's supply." This one goal was partially adopted so that today we have 700 million barrels but only 35 days supply at our much higher consumption rate of 20 million barrels per day. The complete Carter televised speech, including ten fundamental principles, can be found from: Jimmy Carter, *The Proposed Energy Policy*, April 18, 1977, "Vital Speeches of the Day" at **www.pbs.org/amex/carter/film-more/ps_energy**.

There are many think-tanks and administrative agencies, attempting to address the question of how much fossil fuel we have left. Although they do not agree on the exact quantities of fossil-fuel reserves, most agree we are dealing with finite resources that will eventually run short. A partial list later in this chapter summarizes a website search of this subject. Close examination of many of these websites shows a common agreement of about one trillion barrels of oil remaining in the world, enough for 36 years at the 2003 rate of consumption, but not the increased consumption forecast by the EIA. Natural gas reserves show a similar time frame for depletion especially if consumption is increased.

A glaring symbol of excessive energy consumption and how the U.S. public has been kept from the facts in order to provide short-term profits can be found in the ubiquitous sport-utility vehicle (SUV). The book, *High and Mighty*, by Keith Bradsher,[10] primarily emphasizes the poor safety record of these vehicles rather than their dismal-fuel economy. Bradsher also explains the politics that allowed these popular means of personal transportation to be considered as trucks and thus avoid the mandated-fuel economy for automobiles. With over 70 million SUV's produced in the last fifteen years, the increased fuel (the difference between 18 mpg and 28 mpg) they've consumed in that time period (about 0.44 EBBO per year for fifteen years) is about equal to that estimated to be in ANWR and about two and a half times our annual, total car and light truck fuel consumption of 2.6 EBBO (0.44 EBBO/yr x 15 years = 6.6 EBBO). See **www.bts.gov** (Bureau of Transportation) for specifics of U.S. fuel consumption by various vehicles.[11] History may show that the SUV saga was the culminating chapter in the Industrial Age. We almost turned the corner of conservation in the late 70's, but then embarked on another 20 years of steadily increasing fuel consumption.

It does not seem wise for any politician to discuss a subject regarding environment or energy. People don't like to hear disturbing predictions and make personal adjustments. Ten years ago, Al Gore wrote a most prophetic book, *The Earth in Balance*.[12] His message was lost in the prosperous last decade when the SUV and Hummer became kings of the road.

CHAPTER SUMMARY

It has been a most revealing and daunting task to summarize and make understandable something so complex as world and U.S. energy statistics. There is overwhelming information available especially from the Internet. There are so many different organizations manipulating the numbers that one hand may not know what the other is doing. None of the U.S. government information alphabet soup of agencies: DOE, EIA, DOT, and the BTS (Department of Energy, Energy Information Administration and Department of Transportation, Bureau of Transportation Statistics) discuss the energy shortfall of the future.

Other groups are attempting to project a clearer message and sound an alarm, and they do not agree with the optimistic forecasts of the DOE. Several organizations are:

- **The World Watch Institute**[13]

   Quoting their mission statement, "By providing compelling, accessible, and fact based analysis of critical global issues. World Watch informs people around the world about the complex interactions between people, nature, and economics. World Watch focuses on the underlying causes of and practical solutions to the world problems ... etc."

- **The Center for Energy and Environmental Studies**[14] (in conjunction with scientists for a sustainable future and the National Academy of Sciences)

    On May 18, 2001 this prestigious organization published an open (electronic) letter in an attempt to convey a sense of urgency regarding energy. Their website shows this letter and the names of hundreds of distinguished scientists and educators who stand behind the messages relating to the subject of finite-fossil fuels, the folly of drilling in ANWR, the problems with nuclear fuel, and the subject of climate change. Also included is a list of responses to this public letter, including policy statements from President Bush and Vice-President Cheney, showing a concern for the energy issues facing our nation and the world. Hopefully this concern will lead to administrative action.

- **National Renewable Energy Laboratory (NREL)**[15]

    This is one government agency with the honest mission of educating the public and leading us to a sustainable future. But on their home page they specifically state that their "research is paving the way to the hydrogen economy." Perhaps this is a government agency that is guiding our decision makers to a future less dependent on fossil fuels. If they are, where is their input when legislative votes are called regarding energy conservation, permitting cars to be considered as trucks (SUV's), or improving fuel-mileage standards? It seems like either nobody's listening or they serve a cross-purpose to delude the public into believing all is okay and our government is watching out for us. With less than 1% of the U.S. energy supply coming from renewable fuels (other than hydro and biomass), we are not making progress nearly fast enough to prepare for the coming fossil-fuel crisis. I hope the people at the NREL have read the "mitigation" paper commissioned by the National Energy Technology Laboratory (NETL) and noted as ASPO (item 504) in Chapter 2 of this book.

Many additional websites, links, and references provide a variety of energy messages. Following is a partial list of those I have used while researching this book. These websites are in addition to or repeat specific cited sources in other chapters. The most comprehensive and pertinent references are listed first. Many new sites have been added as they have appeared by spring of 2005.

www.hubbertpeak.com—The basic worldwide site concerning oil depletion.

www.peakoil.net—The home site for ASPO (The Association for the Study of Peak Oil and Gas) founded by Colin Campbell. This is the best source for world-energy information.

www.peakoil.ie—The new Irish website address for peakoil.net

www.asponews.org—The newsletter for ASPO is updated monthly.

www.aspo-usa.org—The new USA affiliate. First conference was in Denver, November 10–11, 2005 .

www.peakoilcrisis.com—An excellent starting point for many links.

www.odac-info.org—The Oil Depletion Analysis Center. A very comprehensive U.K. site for current activity regarding peak oil.

www.apolloalliance.org—This is a very aggressive new movement to encourage the public to take control of America's energy future.

www.hubbert.mines.edu—This website is prepared by the M. King Hubbert Center of the Colorado School of Mines in Golden, Colorado. The statistics and interpretation are prepared by U.S. geological experts.

www.greatchange.org—An excellent site with an interjection of humor into a very serious subject.

www.dieoff.com—Another wonderfully diverse and informative source of related information and references.

www.oilcrash.com—A very comprehensive site originating in New Zealand with many links, books, and essays. People the world over are really trying to be heard.

www.postcarbon.org—Founded by environmental philosopher, Julian Darley. Read the essay, "When Markets Fail, America Leaps Off the Gas Cliff Without a Parachute." This site has become arguably the primary U.S. clearing house for all fossil-fuel depletion issues. It is also an excellent source for articles, books, and educational materials like the video, "The End of Suburbia."

www.bartlett.house.gov—Finally a congressperson, Representative Roscoe Bartlett (R) Maryland, recognizes and understands the seriousness of peak oil and is trying to educate his colleagues and the public. See his "House Floor Peak Oil Speech" of 4/20/05 and peak oil conference 9/26/05 on this site.

www.oilcrisis.com—Ron Swenson, founder of the Coming Global Oil Crisis website, provides a wealth of information and a source of reference books and technology.

www.futurereality.org—Economist, Ron Cooke, breaks ranks with his peers and addresses fossil-fuel depletion head on. Also author of Oil, Jihad and Destiny (see book review in Chapter 2).

www.lifeaftertheoilcrash.net—A young lawyer and author, Matt Savinar, is so concerned about the world's future he has refocused his life to alert the public. This site has an endless list of information concerning past, present, and future of oil. His powerful book, *The Oil Age is Over*, is reviewed in Chapter 2.

www.overabarrelbook.com—Retired Texas engineer, Tom Mast, has written a concise book, *Over A Barrel*, to raise public awareness reviewed in Chapter 2.

www.powerswitch.org.uk—A new site from the U.K. "Dedicated to raising awareness and discussion of the impending and permanent decline of cheap oil and gas supply."

www.optimumpopulation.org—Another UK site which focuses on the dire need for population reduction as well as the limitations of all non-fossil energy sources.

www.drydipstick.com—A peak-oil megadirectory for all things related to the subject.

www.peakoilaction.org—People working together to raise awareness. Includes a schedule of upcoming events.

www.theoildrum.com—A daily "beat" (blog) by sophisticated individuals concerned with peak oil.

www.viewfromthepeak.com—All-inclusive site which covers everything.

www.petropeak.com—Another site which is right on.

www.survivingpeakoil.com—Support and ideas for the post-carbon age.

www.communitysolution.org—Advocating strong community cooperation and preparation to cope with the coming crisis.

www.doctorsandpopulation.org—This is a comprehensive site to raise Down Under awareness since Australians have exactly the same modern world problems as the U.S. and Europe.

www.derekjwilson.co.nz—This is the personal site of a wonderful sage and philospher from New Zealand. Read *Where On Earth Are We Going* and *The Five Holocausts*. His wisdom and eloquence abounds.

www.after-oil.co.uk—This is the site for the *Busby Report* referred to later in this book.

www.journeytoforever.org—Another site erroneously promising Utopia from bioenergy. In other words, they infer that short-term incoming solar energy (via bioforms) can supplant millions of years of stored bioenergy in fossil form.

www.wolfatthedoor.org.uk—In their straight-forward terms, "The Beginner's Guide to Oil Depletion."

www.deconsumption.typepad.com—Philosophical musings ad infinitum. Read to your heart's content.

www.energyintel.com—Everything you need to know about world-energy news.

www.321energy.com—Up-to-the-minute trading and market reports concerning energy.

www.ilsr.org—Insititute for Local Self-Reliance. Good links to alternative-energy issues, BUT promises ethanol as a solution with good discussion of the EROEI question (Energy Returned On Energy Invested). (David Morris vs. David Pimental, "Ask Doctor Dave.")

www.renewableenergyworld.com—All things related, including a weekly newsletter.

www.energybulletin.net—For current-market prices and recent publications about every aspect of world economy.

www.spe.org—Society of Petroleum Engineers. The source of the industry party-line estimates 1,270 EBBO left (44.6 years of oil).

www.eia.doe.gov—The very comprehensive, basic U.S. Government site devoted to accumulating statistics regarding U.S. energy production and consumption. This site is frequently referred to in most energy books as the starting point for where we are now and what has transpired so far. Their future forecasts, however, seem to be optimistic extrapolations of the past rather than a realistic understanding of finite-energy resources.

The remaining sites are of more general interest but frequently address specific energy issues:

www.mcintirepublishing.com/energybook—The personal site of Virginia Howe, publisher of this book, *The End of Fossil Energy.*

www.museletter.com—The monthly newsletter authored by Richard Heinberg includes candid reflections about today's troubling times. See Chapter 2 for his book reviews, *The Party's Over* and *Powerdown.*

www.fromthewilderness.com—The other "must read" monthly subscription newsletter published by Michael Ruppert (email or snail mail).

www.simmonsco-intl.com/energyindustry—Matthew R. Simmons, the chairman of an international-investment company, has been a key advisor to the Bush administration, yet his dire warnings do not seem to receive appropriate attention. You can download his lectures on depletion from **www.peakoil.net**. His revealing new book, *Twilight in the Desert* (reviewed in Chapter 2), may well be the most important discourse on the subject of peak oil.

www.climatesolutions.org—Encourages political and popular support for clean energy and action about global warming.

www.populationconnection.org—Focuses on the critical issues of population which are directly interrelated to energy.

www.sustainableusa.org—Renewable-energy technology and conservation strategies.

www.renewables.com—Solar design and renewable-energy products including the solar-powered electric tractor developed by Steve Heckeroth.

www.cleanenergystates.org—The site for CESA (Clean Energy States Alliance). At least 12 states joining together to commit to energy related activity that is not being addressed nationally.

www.wri.org—The site for The World Resources Institute. Another attempt to reach the public regarding sustainability and climate change.

www.aceee.org—The site for American Council for an Energy Efficient Economy.

www.cera.com—Cambridge Energy Research Associates headed by Daniel Yergen, author of *The Prize*. This organization has become identified with the "don't worry, be happy—there's plenty left" movement. One questions their motives and logic (see *The Wall Street Journal* including full-page ads by Chevron, ExxonMobil, etc., Feb. 7, 2006).

www.nrglink.com—*Green Energy News*, an excellent, reasonably priced, weekly electronic-newsletter subscription covering timely energy issues authored by Bruce Mulliken.

www.globalissues.co.uk—This site and the next give similar messages from England.

www.goinggreen.co.uk

www.rmi.org—The site for Rocky Mountain Institute. A strong case for energy efficiency but less convincing about hydrogen and Amory Lovins' hypercar.

www.ieer.org—The site for Institute for Energy and Environmental Research.

www.oilanalytics.org—The alternate site for the Center for Energy and Environmental Studies (CEES).

www.epia.org—The site for European Photovoltaic Industry Association.

www.seia.org—The site for Solar Energy Industries Association.

www.awea.org—The site for American Wind Energy Association.

www.neetc.comwww.directglobalpower.com—Photovoltaic market development.

www.biodieselnow.com—Farm-grown fuel. Where will the food come from?

www.solarsupply.com—Source for renewable-energy products and technology.

www.inri.us—Independent Natural Resources Inc., developer of the Seadog wave pump.

www.syncrude.com—Canadian tar-sand issues from the optimistic industry viewpoint.

www.egroups.com/science/energy—All energy-related subjects.

www.forbes.com/energy—A good reference for current energy-related issues. For instance, see "Rage for SUV's Fuels Oil Import Surge." by Dan Ackman, 5/5/03.

www.climatesolutions.org—Encourages political and popular support for clean energy and action about global warming.

www.earthfuture.com—The name speaks for itself.

www.ucsusa.org—The home site for the Union for Concerned Scientists.

www.nrdc.org—The site for National Resources Defense Council. Energy is only one of many related issues but this is one non-profit that focuses on fuel consumption.

www.growbiointensive.org—This and the next site teach microagriculture solutions

www.pvenergy.com—by Paul Maycock, industry consultant and editor of *Photovoltaic News*.

www.jbenergy.com—by John Berger, environmental author and consultant.

www.soros.org—by philanthropist George Soros. Encouraging transparency of issues needing public knowledge.

www.zapworld.com—Zero Air Pollution vehicles (all electric) are available in all forms.

www.evt.com—Pertaining to electric-vehicles technologies.

www.freerepublic.com   A conservative news site that sometimes explores the news and mysteries of energy.

www.api.org—American Petroleum Institute is a typically feel-good industry source seemingly more concerned with perpetuating the habit of American culture to consume more and ignore the big picture of depletion.

www.whitehouse.gov/energy—The official administrative response alluding to much concern about energy.

www.peakoildebunked.com—Why do people do this? It seems like for every person trying to save civilization, there are two people trying to destroy it.

There are many other cross links and reference sites, which relate to the crisis at our doorstep. Additional new sites seem to appear daily. The problem is all are limited only to internet users and not the general public, which does not use the world wide web and is more influenced by mass media.

We are not going to make it to an acceptable civilized future if the American public does not hear the story of depletion of finite fossil fuels and plan for a future lifestyle without plentiful energy and continued growth.

I repeat again, energy depletion is destined to be the greatest crisis ever to face civilization. Yet, the message is hidden or lost in a blitz of bland media coverage supported by hyperadvertising of high-energy products. Our elected leaders either don't understand or are reluctant to address the subject of energy. Unless we act quickly, we will be fossil-fuel bankrupt in another 20 years with very little left for our children. There will be no possibility for a renaissance of an energy-intensive civilization. The earth's original endowment of fossil energy will be gone. If you question this statement, investigate the references. I contend the more you research the subject the more concerned you will become.

One rare exception to the above statement about uninformed leadership is Congressman Roscoe Bartlett (R) from Maryland's sixth district. Representative Bartlett is one of three scientists in congress and is desparately trying to educate his colleagues about our energy crisis. You can read his reasons for voting "no" on H.R. 6, the Energy Bill, on his website, **www.bartlett.house.gov**. His unique understanding of the imminent energy crisis is a welcome start in Washington—a glimmer of hope. You can also download his one-hour taped program about global peak oil on **www.eande.tv** dated 4/18/05. On September 26th 2005 Congressman Bartlett hosted a peak oil conference in Frederick, Maryland. The transcripts of this public meeting can also be found on his website.

Chapter 4

A Model for an Energy-Sustainable Future

The summary tables on pages 12 and 54 show 86% of U.S. (and world) energy coming from non-renewable fossil fuels. This is akin to spending 100 dollars with 86 dollars coming from savings (capital) and 14 dollars from income. Obviously we're not living within our means. We may continue for a few more years, but there will be little left for our children.

The only hope for a future modern civilization is a fourteen-dollar budget for the fourteen-dollar income. Once we define that model, we need to construct a plan (or bridge) to get there. In addition to these two daunting tasks is the fact that Americans are sharing this singular-bank account with the rest of the world. Other countries, like Germany and Japan, are beginning to plan for the depletion of fossil fuels. The U.S. is lagging behind because of a poorly-informed public. The rapidly modernizing China and India are exacerbating the problem in moving the point of peak oil back to the present time. It's happening now.

Americans are consuming 25% of the world's energy with only 5% of the population. We have the greatest adjustments to make.

Natives on a remote island will have fewer problems in a post-fossil-fuel era (unless climate change from fossil fuels disrupts their balance with nature). Their lifestyle and consumption has changed less from their predecessors. The higher we are on the Hubbert's Peak of energy consumption, the farther we have to descend.

There are a number of well-established, non-fossil-energy sources that make up the remaining 14% of our energy consumption. All, except nuclear, are considered renewable and therefore a basis for an energy-sustainable society. The quantitative contributions from each source, both at the present and proposed for the future, are shown in Chapters 2 and 5.

Hydroelectric—a well-accepted source of energy for years but most favorable sites have already been used for decades and water flow is unpredictable. Also, dam silting makes the renewability of this source questionable over the long term.

Solar—both photovoltaic and non-electric (active or passive), the technology is well established but requires huge financial investments and energy input to scale up to the levels we will need.

Wind—rapidly expanding in many parts of the world but limited to favorable sites. Both solar and wind are sporadic sources of energy which require either energy input for backup or substantial provision for storage.

Bioenergy—only if we live within a "bio-budget" and not divert biomass that should be used for food. Bioenergy production also has to be in equilibrium with other natural resources and fertile-soil maintenance. Cutting more than one tree out of about sixty in one year is not sustainable.

Tidal, Geothermal, etc.—these will continue as localized sources but not in magnitude large enough to affect the overall numbers.

Muscle—The work done by direct human and draft labor sufficed until the last few centuries, but it is so minute compared to what we use now with fossil or renewable sources that it will not be included in the following quantitative design. The intent is to define a future high-technology society without back breaking drudgery.

Nuclear—falls into a contentious zone. It is not a fossil fuel, but like fossil fuels, it is not sustainable because of finite-uranium availability. The uranium ore also requires large quantities of fossil fuel for extraction and shipment. The dangerous record and stigma of nuclear health hazards is well documented. For the remainder of this book, I will continue to use nuclear power at the present level through the Five Percent Plan (as explained in Chapter 5) and into a truly energy-sustainable society. This is necessary because:

1. We will need all the energy we can find to help us bridge the gap to true sustainability.
2. Nuclear energy production will not be expanded. Instead, existing facilities will be allowed to run their course.
3. It is a substantial part of energy supply in other parts of the world and when used properly is gaining some respect.
4. The problems of waste disposal and public perception appear to be improving as remaining facilities are carefully controlled. This is only an opinion. At least things do not seem to be getting worse in the mature years of this energy source.
5. Improvements in technology may allow us to continue using some forms of nuclear energy beyond the demise of fossil fuels. We should continue its use and development but on a very guarded basis.

With only the energy available from these non-fossil sources, it is paramount for survival that we design a new, low-energy lifestyle. Then we must plan on how to get to this new state with the remaining fossil energy we have left in the bank before it is frivolously used. If we lose this opportunity to change now, our chances for a modern civilization will be lost forever.

A vision of a modern lifestyle using non-fossil energy, rather than reverting to a hunter/gatherer/agricultural subsistence and without significant breakthroughs like fusion energy is as follows:

1. Continue using centralized generating systems and utility grids we have now. This will allow a gradual transition to the new fossil-fuel-free system by blending together the present era with the future.
2. Start to immediately build decentralized (distributed) solar energy systems (both electrical and heating) in all new and

existing homes. Every home becomes an independent power plant. We have the technology now; our ancestors did not.

3. Drastically downscale all travel and movement of goods with the ultimate goal of electric mass transportation. Small electric vehicles will provide for individual needs. It's either that, ride the bike, walk, or stay home. Transportation energy is especially vulnerable because about 90% comes from oil alone. In the U.S., one-half of our 20 million barrels (400 million gallons) consumed each day is used just for gasoline. This will soon end no matter what we do.

4. Save precious remaining fossil fuels for agricultural, military, municipal fuel, important mass transportation, and emergency uses. If we start immediately to conserve these fuels, we might buy enough time to invent new high-energy sources while we still have the infrastructure of a functioning society. At least we could save enough fossil energy to ration over an additional hundred years. (See Chapter 5, "The Five Percent Plan.")

5. Our remaining finite supply of fossil fuels must also be available to make the transition to sustainability. This will make possible the infrastructure we need to make solar cells, windmill blades, and continued research and development of new-energy sources.

Before describing the details of the plan, let me reiterate, I am an engineer faced with a problem and proposing a solution. If you question or disagree with the premises and conclusions, verify the presented facts. Search for other opinions just as you would if faced with major surgery. Remember, we collectively have an undeniable terminal illness. Our leaders (except for President Carter) have been guilty of malpractice for years for not telling us the truth when we could still do something. The only hope is to do something drastic, soon, before it's too late. I am only a messenger presenting well-documented data and logical conclusions.

There are three basic parts in the proposed future sustainable model:

1. THE PRIVATE RESIDENCE (PR)

Let's start with the private home. This decentralized (distributed) energy source will have to supply most of the energy for both the dwelling and a personal transportation vehicle (PTV) or multipurpose solar-powered electric utility vehicle (SPUV).

Each home becomes a small free-enterprise electricity producer of about 4 kw peak power capacity. This requires 40 square meters (435 square feet or a typical roof side) of photovoltaic panels. The cost in 2003 dollars would be about $20,000, enough for a basic 2-car garage. At this cost, the 87 billion dollars just appropriated for Iraq would take care of $4\frac{1}{2}$ million homes. Half (2 kw) of this capacity would suffice for adequate requirements in parts of the country with more sunlight. By connecting with a utility interactive (UI) grid where power can flow, to or from the central distribution facility (CDF), many homes smooth the storage for others, interconnect with centralized energy sources, and reduce the need for batteries. Residential power bought or sold from the centralized grid would be metered and priced high enough to encourage conservation and maximize individual output.

Typical energy usage from a 4 kw peak-power system (2 kw in the sunny southwest) would yield, at 14% capacity factor, about 400 kwh per month (4800 kwh/yr) even in the cloudy northeast. This 400 kwh per month budget could allow a typical household consumption budget as follows:

200 kwh	for personal transportation (PTV) or multipurpose solar-powered vehicle (SPUV) used for both transportation and home agriculture (this could be supplemented by on-board solar panels)
85 kwh	hot-water heating augmenting by passive solar (good for one shower or bath every 2 days)
15 kwh	high efficiency (12ft.^{3}) full-size refrigerator
30 kwh	(2) small cooking burners
8 kwh	(1) well-insulated oven
12 kwh	TV, computer, and communications

8 kwh	washing machine
12 kwh	lighting
12 kwh	water pump
12 kwh	miscellaneous power tools and efficiency losses
6 kwh	toaster and vacuum
400 kwh	Total/month

In June 1997, the Clinton administration proposed the "Million Solar Homes Initiative." Now, six years later, little has happened, and time is running out. (One bright spot... as of May 2005, California State Senate overwhelmingly passed SB 1, "The Million Solar Roofs Initiative.")

With the proposed 4 kw system on 40 million homes (almost all the private residences in the U.S. today), the equivalent fossil-fuel power-plant production, at 35% efficiency would only be 0.3 EBBO. (See Appendix 1 for detailed calculations.) This proposed individual residence (PR) electrical-energy plan supplies far less than our present wasteful energy consumption, but for Third World societies or our grandparents, such energy availability and consumption would be positively extravagant.

Remember, this decentralized distributed-energy plan only supplies our electrical needs for personal transportation (and/or micro-farming), a small bit of hot water for dish washing and about 30 quick showers, and minimal other residential requirements. All of the home heating and most hot-water requirements would have to come from active or passive (thermal) solar heating, well-established technology used today throughout the world. All new homes should be built on a smaller scale and with maximum thermal efficiency. Existing homes could be retrofitted and reconfigured so the actual living area reduces to a warm core part of the home for the cold winter months. Assuming the non-photovoltaic solar heating installed in each of the required 40 million homes for domestic hot water and some domestic heating is equal to the 4800 kwh/yr. of photovoltaic energy, the total would double the decentralized supply output to 0.6 EBBO annually (0.3 EBBO photovoltaic and 0.3 EBBO thermal).

This total amount of photovoltaic-electric, active, and passive-solar energy generated from 40 million residences (0.6 EBBO) is the SAME ORDER OF MAGNITUDE as the extra fossil fuel consumed in one year driving SUV's or light trucks instead of commonly available 35 mpg cars (a difference of 0.44 EBBO). Another way of considering 0.6 EBBO is, this is equivalent to about 65 days of gasoline consumption at the present rate.

The battery storage for each personal residence (PR) depends on individual situations and the dependability of the intertie central-distribution facility (CDF). Also, the on-board batteries on the personal-transportation or electric-utility vehicle could supplement the stationary battery pack in each home. As miniscule as it may seem, each of the 800 kwh/month residential systems (400 kwh photovoltaic plus 400 kwh passive) is equivalent to the human energy of 67 people working for one month of forty-hour weeks (160 hours) at 75 watts, without the need for food or shelter as required for slaves or servant.

The other renewable fuel for domestic heating in colder climates is the old bioenergy standby, wood. We can only use this sparingly as it is very polluting, and it would take at least an acre of woodlot to continuously provide one cord of wood every year to heat a small, efficient home. The old Yankee rule of thumb is five acres of woodlot for yearly sustainable heating of a typical, drafty farmhouse. Any faster rate is not sustainable, and we would only be using our kid's resources. Wood heat sounds great in Maine but can be devastating if consumption exceeds sustainable growth. This can be seen in Third World countries where people walk miles each day to gather wood for heating and cooking. Besides, wood smoke is very harmful. Wood burning is banned in many areas of the country due to congestion and/or poor air flow (see *Strangely Like War* review on p. 31.)

An additional intriguing use of PR electricity production could be for a personal-agricultural tractor. In the spring and summer an excess of solar or photovoltaic energy from each home could be combined with on-board solar power and used to power a small tractor of between five and fifteen horse power. A number of attachments could be used such as a sickle bar mower, rotary tiller, traditional plow, or wagon. It could also be used to power an electric chain saw, splitter, log winch, or other portable tools. It is also conceivable that the personal-agricultural-tractor function could be combined with the personal-transportation vehicle in a single solar-powered, electric utility vehicle (SPUV). In direct sunlight, the proposed residential photovoltaic array (equivalent to five horse power) could almost supply the tractor with full power on a one to one time basis with incoming solar energy especially if an additional photovoltaic canopy is mounted on the electric vehicle. (See Appendix 7 for considerable additional input regarding the SPUV after a year's experiment with the working model shown on the back cover.)

If we spend time growing food instead of mowing lawns, the number one crop in the U.S. with 32 million acres, we will be much less dependent on centralized agriculture and food distribution. Home food production will have to become a way of life.

---

2. THE CENTRALIZED DISTRIBUTION FACILITY (CDF)

The central connection for the new energy system will be a power distribution facility preferably located at or near an existing public utility plant. The purpose of the CDF is to mix all energy inputs and outputs for a surrounding region and smooth the demand and supply so net-energy flow is neutral. The CDF could also have pumped water or low-pressure hydrogen to smooth the peaks and valleys of supply and demand. This would minimize the need for batteries in each PR. Publicly owned CDFs would also include large-scale photovoltaic production equal to the total residential (distributed) output of 0.6 EBBO bringing the total U.S. solar contribution by 2050 to 1.2

EBBO/yr. Extensive wind energy farms could supply the CDF with another 0.3 EBBO/yr. This is twice the wind energy projected by the Energy Information Administration (EIA) in their high renewable National Energy Modeling System (NEMS) for the year 2025. With my plan the total electricity produced by renewable solar and wind would be 0.6 (CDF) plus 0.3 (PR), plus 0.3 (wind), or 1.2 EBBO/yr. This is just 20% of our 2001 electrical usage of 5.96 EBBO/yr. This would only be possible if we immediately begin this massive energy transition.

The large financial investment for each CDF would come from a stockholder corporation. The energy flow would be closely controlled and supplemented in a manner compatible with government mandates considering the availability of diminishing fossil fuels. Close monitoring and regulation will be required to control the pace at which contemporary fossil fuels are rationed and phased out. We need national guidance for this huge task at hand. Governmental regulation must have the long-term survival of our society as top priority. Short-term market forces and profits will only keep us on the dead-end track we're now on.

Our only hope is for an informed constituency, to demand equitable laws and fuel rationing for everyone. This is a lot to ask, but it must happen.

---

3. THE PERSONAL TRANSPORTATION VEHICLE (PTV) and THE SOLAR-POWERED UTILITY VEHICLE (SPUV)

Automobiles and airplanes are the highly-visible symbols of the fossil-energy age. Only liquid hydrocarbon fuels can provide the necessary high energy to low weight and volume ratios required for rapid personal movement. Over 90% of our transportation energy is supplied by oil. A centralized coal plant, nuclear, or hydro cannot provide the lightweight-energy source we need to carry with us on a trip. Extension cords don't go very far, and it takes at least fifty times more weight in lead-acid batteries to provide the same energy storage as liquid fossil fuels. Arguably, synthetic fuels or steam energy could be made from coal, tar sands, or biomass. These would be a temporary answer only until we totally run out of these dirty fuels or deplete our forests and croplands. Biomass energy should be reserved for personal-energy requirements (food), topsoil maintenence, and sustainable wood heating.

The only other known way to concentrate enough energy for rapid personal transportation is to use renewable-energy sources to make pure hydrogen from the electrolysis of water. This liquid hydrogen would have to be carried on board at very high pressure or in a matrix of exotic-metal hydrides. Either way would require a very large container or extremely thick walls. As needed, the hydrogen could be converted to electricity in a complicated fuel cell or be used as an internal-combustion fuel. As stated earlier, billions of dollars have already been spent on this technology with little commercial success to date. The best possible place for hydrogen in a realistic distributed-energy model may be as a temporary energy-storage system at the CDF. Excess electricity flowing into the CDF could be converted by electrolysis to hydrogen. The short-term stored energy could then be converted back to electricity during periods when demands exceed supply just like pumped water storage. The popular subject of hydrogen is discussed further in other chapters.

At this time, there does not appear to be any established technology to satisfy the energy requirements for high-speed personal transportation in a non-fossil-fuel society other than biomass, which requires considerable land area and input energy.

Mass transit could revert to rail systems with electricity coming from the CDF. This would be similar to today's subways and the electric trolleys of the early 20th century. Water travel will have to go back to sail power possibly augmented by solar electricity. We will have to stop moving ourselves and use our new communication technologies for conveyance of messages and visual images. Shipment of goods will have to be reduced to an absolute minimum.

As fossil fuels become more scarce, personal transportation (other than on a domestic animal) will revert to walking, bicycles, or possibly a personal-transportation vehicle (PTV), which is battery powered from energy generated from the (CDF), the private residence (PR), or an onboard solar panel. This proposal may at first seem unrealistic, but the technology is available in golf carts and electric vehicles already on the market. There have been many attempts to market electric-powered vehicles (see **www.zapworld.com**), but commercial success so far has been limited because of overwhelming competition from fossil fuels and danger from the size and speed of todays cars and trucks as well as illegality on public roads.

A typical personal-transportation vehicle (PTV) with the following specifications can be built with today's technology:

Gross weight	1,250 pounds
Dry vehicle weight	450 pounds (fiberglass and aluminum) (550 with on-board PV panels)
Batteries	500 pounds (approx. 8 kwh capacity deep cycle)
1 passenger and load	200 pounds (not very much! There could be room for a 2nd passenger and possibly a child.)
Max continuous power	3 kw. Typical of GE model D29EO DC golf-cart motor (65 lbs.)
Voltage	48 v DC
Max current	78 amps, (continuous 50 amps)
Coefficient of drag (C_D)	less than 0.3 (Readily achievable using a shaped fiberglass body.)
Frontal area (A)	less than 12ft^2
Top speed	over 50 mph. (No problem, could reach 60 mph on the level.)

(continued)

Hill climbing and acceleration will be adequate and will depend on the type of transmission and motor controller used.

Range   3 hours or up to about 100 miles.

(with regenerative braking and efficient wheel hub motors)

This range might be stretched to above 4 hours and 160 miles at 40 mph since it takes much less force to push the vehicle at slower speeds. The force to overcome wind resistance is proportional to speed squared. One-half the speed takes one-fourth the force and energy. That's why we will have to move much more slowly in the future as explained in Appendix 6, Energy 101.

Photo courtesy of John Snyder www.johnsnyder.biz

The above photo shows a 48 volt, solar-powered golf cart, which works great, is fun to drive up to 20 mph, and will run for four to six hours without additional sunlight while using an 8 kwh battery pack. By summer of 2006 I hope to have my 1962 electric MG Midget on the road. This vehicle should exceed 120 miles at 40 mph without recharging from the on-board 72 volt PV array.

An alternative to a light-weight PTV could be a larger solar-powered utility vehicle (SPUV) that would not only provide tractor power for micro-farming but could double as a 20-mph PTV on the road. This dual-purpose vehicle might have the following design and performance specifications:

Gross weight	2500 lbs.
Dry vehicle weight	1400 lbs.
Batteries	1100 lbs. (approx. 17 kwh capacity deep cycle) plus another 500 lbs. on a trailer
Continuous power	5 kw or 7 hp (typical of advanced X91-4003 model, 60 lb. motor)
Voltage	72 to 108 v DC
Max. current	115 amps continuous
Top speed	20 mph.
Range	25–50 miles (with photovoltaic canopy or approximately two to three hours of land plowing, the toughest agricultural requirement)

This SPUV would be similar in size to a 1950's vintage Farmall Cub. By adding 140 square feet of photovoltaic canopy (70 sq.ft. above the tractor and an equal amount above a trailer) an additional 1400 watts (2.0 Hp) could be realized in direct sunlight. The electric SPUV concept makes sense because extra battery weight is beneficial for traction and there is more sunlight in the spring and summer months. An example of such a vehicle has been built by the author and is shown on the back cover (see Appendix 7 for more details). Solar-powered vehicles could augment the residential PV system in the winter months and in the summer months surplus residential PV power could increase the charging rate of the solar-powered vehicles.

These proposed concepts for transportation and farming sound feeble and should not be compared to 150 Hp cars and 50 Hp tractors but they offer the only chance to continue a modern civilization. In the long-term, large-scale commercial farming will eventually have to partially revert to oxen, horses, or mules because the manure (also called black gold at one time) is essential to maintain soil productivity. A farmer cannot continue to export energy in the form of food (or biofuel) without returning to the land what is removed. Commercial fertilizer and inexpensive fossil fuels will no longer be available. Eventually, population will have to significantly decline and disperse back to many small farms. A comprehensive plan for energy reduction as proposed in Chapter 5 is the only way we can make it in the future.

The concept of a solar-powered and a wind-powered society is not new. Several European countries have committed to substantial wind energy contribution to electrical needs by 2020. Dr. Hermann Scheer is a member of parliament for the German government and winner of several prizes for his work on renewable energy. His comprehensive book on the subject is titled *The Solar Economy, Renewable Energy for a Sustainable Future*[1] and is now available in English translation. In addition to direct solar and wind, Dr. Scheer relies heavily on biomass as a major substitute for fossil energy. Unfortunately, he does not quantitatively explain the sources of energy to produce the biomass, the loss of food by diverting bioenergy to biofuel, and the problems of water and soil depletion.

As so well explained by David Pimental in his book, *Food, Energy, and Society*, reviewed in Chapter 2, in a fossil-fuel-free civilization, all energy will have to come directly from concurrent solar radiation or from biomass. This must also include the energy to reproduce the Industrial Age hardware necessary for long-term sustainability. The head start we have now, due to the carry over of remaining fossil energy, will eventually be used up. Our descendents will not only have to obtain renewable, non-fossil energy for food and all other needs, they will also have to rely on incoming solar power to manufacture new solar cells, batteries, copper wire, structural steel, aluminum, and all the other components and materials normal to an industrialized society.

This challenge may seem overwhelming and barely conceivable, but it can be done in theory. For instance, a photovoltaic-solar system that could last 50 years requires only one or two years of returned energy for replacement. The following table quantifies the weight of various materials that can be raised to melting temperature (specific heat) and then melted (heat of fusion) with just one 4 kw (the size proposed for each residence) photovoltaic system providing 4800 kwh (16 million BTU) per year at 50% efficiency for one year:

Lead batteries)	264,000 pounds (enough for about 4,000
Copper	29,304 pounds (enough for 400 electric cars, tractors, or homes)
Steel	22,408 pounds (enough for 200 vehicles or 20 transmission towers)
Aluminum	23,120 pounds (a substitute for steel)
Glass	5,200 pounds (as needed)

Obviously this type of energy usage and capital investment could only be done at centralized facilities. Each community center would have to have battery-recycling capacity. Also, transportation and the energy for ore extraction will have to be provided by centralized solar, wind, and hydro using batteries or hydrogen for storage.

Eventually, even the ore for each material will become so depleted that all subsequent material requirements will have to come from recycling. By that time, we will experience population reduction and drift slowly backwards from any semblance of an industrialized modern civilization. It may be that the 200 year, fossil-fuel consumption age is just a small blip in a much larger ten to twenty-thousand year finite-metal-resource epoch that spans the copper, aluminum, and iron ages of civilization.

Taking this line of reasoning further, it could be argued that it is impossible for an advanced civilization to exist anywhere in the universe for longer than a few thousand years. The very unique circumstances and time required to concentrate metals and energy on a singular planet cannot provide enough resources to last longer than a few years in the cosmological time frame. The time frame for space-travel evolution may, in reality, be impossible.

This is not a unique concept (see *The Olduvai Theory: Sliding Towards a Post-Industrial Stone Age*[3]). This philosophy of previously stored, time-limited resources may answer some very perplexing questions:

- Are we alone in the universe?
- How did we get to the standard of living we enjoy in our modern industrialized world?
- What will be the future of our species?
- What life on earth will our immediate and long-term descendants experience?
- Can we control our destiny?
- What will be the time frame for our civilization?
- How does our species fit into the long-term picture of biological evolution?
- Would it be better for fewer people to enjoy a longer Industrial Age or more people to shorten it?
- Is there a religious interpretation for the depletion of finite resources, both long-term for metals and short-term for fuels?

Finally, it is worth noting that there are two excellent contemporary and contrasting examples of post-fossil-fuel, industrialized societies. When Russian oil exports to North Korea and Cuba ceased, following the collapse of the Soviet Union, both countries were faced with a serious life-threatening shortfall of energy. North Korea has done little but experience infrastructure collapse and starvation. Cuba, on the other hand, has made an about-face to decentralized microagriculture and has become a model of agricultural reform in the developing world. For instance, urban gardens alone grow half the produce consumed in Havana and 215 grams (about one-half pound) of vegetables per day per person for the entire population. (For in-depth discussion see *Drawing Lessons from Experience: The Agricultural Crisis in North Korea and Cuba* by Dale Allen Pfeiffer.[4]) It seems like the total Industrialized World is poised today at a similar fork in the road.

Chapter 5

The Five Percent Plan to Energy Sustainability

After 100 years of plentiful energy, we find ourselves perched precariously on top of "Hubbert's Peak." In Chapter 4, we envisioned a fossil-fuel-free lifestyle and a future with much lower energy. The challenge for our high-flying society is how to get there before we completely run out of the temporary, high-energy support system on which we've built our present levels of population and consumption. This may not be the best scenario because it infers that all we have to do is coast back down to a sustainable future. In reality, we'll need all the fossil energy we have to make the transition to soften the landing and still leave a small amount left for absolute necessities.

An example of our present plight might be a team of astronauts on the moon. From this precarious position, they must get back to a basic sustainable environment before their artificial, finite life-support system runs out. If their previously stored energy supply is depleted, they will never make it.

Like astronauts or occupants of a life boat, we must share the challenges and resource rationing of the return trip as provisions become scarce. Astronauts do not share provisions on an ability-to-pay or grab-what-you-can basis. It seems preposterous that intelligent, informed human beings resist working together to survive, but that's exactly what is beginning to happen as we squabble like typical animals over dwindling resources.[1,2,3]

Ernest Shackleton provided the leadership necessary to save all of his men when his ship, the *Endurance*, broke up in the Antarctic. Noah built an arc. We need leadership.

Many reputable scholars and scientists have been warning for years of the pending collapse of our high-energy civilization. How much time we have left to design and cross the bridge is not exactly clear, but it most certainly depends on how soon we get started. There appear to be two positions on the subject. One group predicts an inevitable decline of high-energy civilization in the next 10 to 20 years. Referenced websites in Chapter 3 link to many highly-qualified and concerned authors. Their books and papers have titles like *The Olduvai Theory, Sliding towards a Post-Industrial Stone Age* by Richard C. Duncan (the Olduvai Gorge is a well know prehistoric site in Africa where humans originated),[4] *Energy and Human Evolution* by Dr. David Price,[5] the recently issued *The Party's Over* by Richard Heinberg,[6] *The Oil Crash and You* by Bruce Thompson,[7] and *Oil Depletion in the U.S. and the World* by Seppo A. Korpela.[8]

The second, optimistic camp seems to be dominated by government supported agencies and related research groups as is described elsewhere in this and other books. Imbedded in this second group are shadows of the business world and world-energy corporations. Although some of these companies, such as British Petroleum and Shell, have websites delineating fossil-fuel reserves and consumption rates, none (until Chevron in 2005) seem to want to send a clear message of hard times ahead. I think of these as feel-good websites versus the gloom and doom voices of the first group. It is difficult to understand the different approaches to this subject because the above mentioned oil companies do agree, quantitatively, with the gloom and doom prophets of the first group. But there has been a tradition among economists and oil people that since the reserve to production ratio (R/P) of oil, natural gas, or coal has remained somewhat constant in the past that all is okay. Short-term interests and profits are satisfied. This is misleading. A better indicator of future production is the ratio of new discoveries to present consumption. This number peaked in the 80's and has been declining for years. We are now using oil at a rate four-times higher than we are finding it. There are also real suspicions about "transparency" reporting and number manipulation by both nationalized and public oil companies. This is especially true of the Middle Eastern countries lead by Saudi Arabia and discussed by Matt Simmons in his book, *Twilight In The Desert*. In late 2005 this book became a best seller among business books.

THE SEARCH FOR SUSTAINABILITY

To design the path (bridge) to sustainability, I will continue to use equivalent billion barrels of oil (EBBO) as the fundamental unit of consumption for all finite fossil fuels as well as the nuclear and renewable energy sources which will become the basis for our future economy. Natural gas is our next favorite fuel, and these reserves are not much better than oil. As we've seen, the world's oil production is at or near its peak in 2005. Natural gas is definitely becoming harder to find and production will fall soon after oil. There are more sources of coal, but its use is very environmentally destructive. It also takes considerable fossil-fuel energy to remove it from the earth. It has been projected that by the year 2040, it will take more energy to mine and process coal than the coal can provide.[9]

To design the plan to sustainability, we will consider natural gas and coal to have the same finite-reserve life as oil. As an engineer, I've learned to design on the conservative side. Engineers like to talk in terms of safety factors to allow for unexpected pitfalls and fuzzy numbers. For a subject as serious as the future of

Total U.S. Energy Consumption Showing the
Five Percent Plan to Energy Sustainability

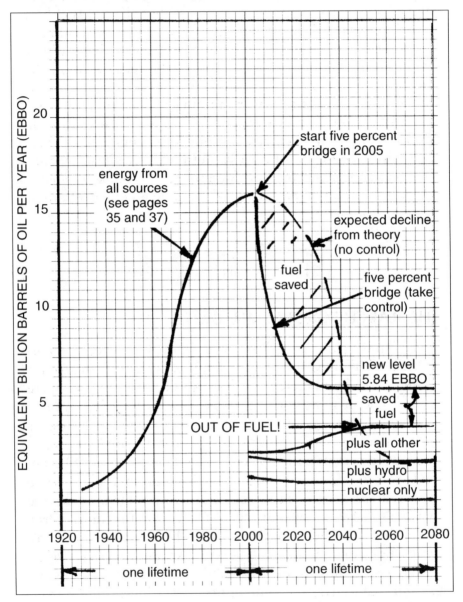

modern civilization, it seems prudent to err at least in our favor. As oil production declines, natural gas and coal will also peak and deplete more quickly. It has been argued that heavy (non-conventional) oil and tar-sands oil will keep us going a few more years. We should not depend on these inferior sources for our future. Besides, it is accepted fact that we will never be able to extract all of the remaining oil.

In Chapter 2 and 3, I summarize annual U.S. energy consumption to date, in terms of equivalent billion barrels of oil (EBBO) used each year. The path to the future, as described in Chapter 4, starts with this same bell-shaped probability curve. I focus on U.S. consumption for a number of reasons:

- Other countries such as Germany, Denmark, and Japan are already ahead of the U.S. in their quest for sustainability. Also, some countries use different mixes of energy to make up their total consumption.

- We live in a global economy using global resources. The U.S. uses far more total energy, twice the energy per capita than any other developed country. We are the bad guys using more than our share. We have the largest correction to make. The world looks to the U.S. for leadership, but they're not seeing it in terms of energy conservation and rapid deployment of renewable energy. Even the Utopian fossil-fuel, free-energy model envisioned in Chapter 4 provides more energy for the U.S. than the undeveloped Third World uses today and considerably more than our great grandparents had without fossil fuels.

- If we take the initiative in this endeavor, the world will think better of us than for our reputation of laying claim to remaining reserves and conspicuously consuming the world's resources.

The solid line on the right side of the preceding curve is our direction following a controlled Five Percent Plan. The broken line is the expected decline in consumption (especially oil and natural gas) if we do nothing. Reaching sustainability hinges on four significant commitments. Without all four, all bets are off. We can only expect to slide into chaos, anarchy, and starvation.

89

1. CONSERVATION, THE FIVE PERCENT PLAN

The curtailment (conservation) of fossil-fuel consumption is the focal point of the second half of this book. We must start immediately to reduce our oil, natural gas, and coal consumption by at least five percent per year. This will be fairly easy for us because of today's excessive waste. In addition, we should realize improvements in our overall well-being. I will expand on the details of the changes we have to make in Chapter 6. For our and our children's future, we need to start on the Five Percent Plan now. A close look at the numbers shows why this is true. (The quantitative annual reduction for the Five Percent Plan is summarized in Appendix 4.) As this book is revised in 2005, we have used an additional 55 billion barrels since 2003. That's why this is now our last chance for sustainability.

2. NEGATIVE POPULATION GROWTH (NPG)

We cannot continue to grow in population. As we run short of fossil fuels, especially for producing nitrogen fertilizer and the energy-intensive agriculture needed to grow and distribute food, we will not be able to feed more people or even the numbers we have now. It takes about one EBBO of energy just to feed the current U.S. population of 280 million people. We must begin to move toward negative population growth. The average number of children per couple must be less than 1.5 if long-term population is to peak and begin to decline. One child per couple should be the goal. Extra children consume more resources. Quality of life, per capita energy, and food supply all suffer accordingly. This is a tough subject, but if we don't take control of the numbers, simple mathematics will be our demise. Today, the majority of the world already suffers from the simple problem of overpopulation. There are too many hungry bodies and not enough food. This is occurring even with an overflow of fossil energy, which filters its way into the Third World but will not be there in the future. Hard times are coming because we did not stop multiplying when resources were plentiful. Ultimately, our population will have to decrease by fifty percent or almost proportional to our decreased-energy usage. The low-energy world of the future will not be able to provide for all of the children born today.

3. GO SOLAR

The proposed sustainable-energy plan hinges on a total commitment to solar electricity and heating, especially on a residential basis. The plan is outlined in Chapter 4, "A Model for an Energy Sustainable Future." This plan proposes that 40 million individual residences (about one for every seven people of today's 280 million U.S. population) to have solar systems by 2050. Chapter 4 explains how and why decentralized solar-energy sources will also have to provide energy for most personal transportation as well as substantial local and private food production.

Each of the 40 million residences will produce an average of 4800 kwh per year of electricity and an equal amount of solar heating. This decentralized residential-energy capacity combined with an equal amount from centralized solar-electricity power plants will provide us with the equivalent of 1.2 billion barrels of oil (EBBO) per year. (See Appendix 1 for the math.)

4. STAY PUT OR TRAVEL SLOWLY

The short petroleum age provided the energy to make possible the excessive distances we travel today. This can be seen in the work equals force times distance equation (Energy = Work = Force x Distance). All frivolous movement of people and things will soon end. As stated earlier, any speed above 20 mph begins to encounter wind resistance as a major factor. Most of the energy of today's travel is expended to overcome the force required to churn and heat surrounding air when we exceed 20 mph.

Drastically reduced travel and speed are not easy pills to swallow, but we have no choice. The math and dwindling high-energy resources will make these changes inevitable. If we control the coming change, we will still have a minimal amount of fossil fuel left for national defense, municipal needs, and emergency travel. Life won't be that bad, our great grandparents survived without moving too fast, exercised through work, and met their neighbors.

THE LOW ENERGY GOAL

The following table starts with the present U.S. energy consumption as shown in Chapter 3 and updated to 2005 in Chapter 2. These numbers are then projected forward to what they should be by 2050 and are in EBBO consumed annually.

(All units are EBBO)	Present U.S. Total Consumption (2001)	Future Energy Consumption (2050)
Finite fossil fuels	13.28 (oil 6.05, natural gas 3.69 coal 3.55)	2.0 from 200 EBBO saved and spread over 100 years using the Five Percent Plan
Hydroelectric	0.379/ 2001, (0.622/1996 with better flow)	0.7 all possible sites maximized. Average flow/year
Nuclear	1.20	1.2 we will keep nuclear at the present level but continue to clean up extraction and waste disposal processes (see discussion in Chapter 2)
Wood	0.347	0.3 reduced to sustain resources and because of decreased fossil fuels available for harvesting and transportation
Biowaste	0.088	0.1 continue at present level, but we must have sustainability, including wood, or we'll look like Easter Island

(continued)

(All units are EBBO)	Present U.S. Total Consumption (2001)	Future Energy Consumption (2050)
Ethanol (and biofuels)	0.023	0 deleted. No fossil fuels will be available to produce ethanol for fuel. All agricultural output required for food. Biofuels are eliminated because of excessive energy input and required arable land.
Geothermal	0.05	0.05 continue at present level due to rarity of sites
Solar	0.01	1.20 increased more than 100 times as the cornerstone of our future energy needs (see Appendix 1, Chapter 2, and Chapter 4)
Wind	0.01 (0.018 in 2003 significant progress made here)	0.30 30 times the present rate. There is considerable additional potential here. The proposed rate is almost twice the NEMS (National Energy Modelling System) high renewable case for 2025.
Total	15.5 (increase to 15.75 in 2003)	5.85

The future energy consumption goal to be reached by 2050 is about 6 EBBO per year. Of this available energy, 36% (2 EBBO) is carried over from fossil fuels conserved using the Five Percent Plan.

This reduced level could keep us going for another 100 years. If we continue to squander our fossil fuels, as we do now, we will have no chance to bridge the coming crisis and "mitigate" (that new DOE, Hirsch-Report buzzword) the landing from Hubbert's Peak (see ASPO News, item 504 in Chapter 2).

The lower-energy goal in 2050 is about one-third of our present level of waste and consumption. It would provide a very tolerable lifestyle for our descendents, at least until the middle of the next century. We would still have far more energy than our ancestors who only had muscle power, wood, and draft animals but far fewer people with no airplanes and no cars.

If we can buy time for an orderly transition, we could still have the possibility to develop new energy sources such as nuclear fusion or gas hydrates. As explained earlier, the "don't worry, be happy" party line is we've always managed to discover or invent new energy sources on a timely basis or devise improved methods to extract more from existing reserves. Experts are now telling us loud and clear, oil and natural gas supplies will significantly decline in the next twenty years. Coal appears to be more abundant but in deteriorating sites of accessibility and quality. As explained earlier, more energy will be required for extraction, and the environmental record for coal is terrible.

The curves and numbers in this book show the energy dilemma we're in, especially in the U.S. By reducing our demand on world resources by 200 EBBO over the next 40 years, we could spread the remaining fossil fuels over the following 100 years from about 2050 to 2150. This would give us at least one EBBO per year to provide food for our population (see Appendix 2 for the calculation). The other one EBBO saved per year using The Five Percent Plan would provide for a continuation of centralized governmental infrastructure, national security, pharmaceuticals, lubricants, mass transportation, manufacturing renewable-energy equipment, all other industry requirements, plastics, etc. In magnitude, the annual conserved fossil fuel energy from the Five Percent Plan (2 EBBO for each of 100 years starting in 2050) is almost twice what we expect from 4 kw solar systems installed on 40 million homes (plus an equal amount of centralized solar-energy production). This may seem futile as we continue to run out of fossil fuel, but urgent investment and familiarity with renewables can lead us into a truly sustainable energy future. There is a strong movement happening which shows that this huge change in technology and lifestyle could produce millions of new jobs.[10]

We can deny or argue with the cited references, premises used, and scenarios offered throughout this book but consider the following:

- We are clearly depleting and running out of fossil fuels. At our present consumption rate, we will begin a serious downhill slide in the next three to five years, a time when growing world population needs more energy for food and strives for more luxuries. Rising demands and dwindling resources will result in skyrocketing prices and inflation.

- The U.S. uses about twice as much energy per capita than any other country. We must make the greatest reductions in consumption to reach a new low-energy society. If the U.S. takes the lead in downsizing consumption, we could lead the rest of the world in a coordinated effort. If we just reduced our per capita consumption fifty percent, to the level of the remaining western industrialized world we would be a world leader in conservation.

- There are no panaceas or alternatives. It is too late to rely on unproven technologies (see Chapter 2).

- The proposed Five Percent Plan takes us only into the middle of the next century. By then, we will be void of most fossil fuels as well as most other finite resources. Hopefully the extra 100 years gained by conservation will give us the time to adjust at an acceptable rate.

- By 2150, U.S. population should also decrease to about 70 million people (which is about one-fourth of our present 280 million) in order to continue a truly sustainable civilization. The rest of world population will also have to decline at the same rate either by design or the cruel forces of resource depletion as is already happening in much of the Third World.

Our planet will never again be able to support a high population, industrialized modern civilization at the level we now enjoy. All of the high-energy fossil fuels, which took millions of years to accumulate, will have been used up in a couple of hundred years. There will be little left for future inhabitants and no high plateau of technology to provide the luxuries we take for granted.

- It is understandably difficult for us to notice the transition from increasing to decreasing energy availability. The change from a positive to a negative trend is taking place over several decades and is blurred by many other factors such as price, economics, wars, misinformation, and geopolitics. This transition is called an inflection and can be compared to a summer or winter solstice. It is difficult to notice the day by day change in sunlight and weather, but we know it's going to happen and plan accordingly. Winter clothes or bathing suits appear in stores long before the appropriate season arrives. Why don't we plan the same way? We know that overhunting and overfishing are counter-productive so we enact laws to prevent depletion. We need to do the same for energy.

- The only hope to maintain the extremely high level of consumption we have now will be to harness a perpetual, clean form of energy like nuclear fusion, which in turn could provide copious amounts of electricity and hydrogen with all its attendant problems. The sophistication, energy, and time for this type of research and development cannot be developed in a subsistence society living in huts.

- The only non-fossil energies now are hydro, nuclear fission, bio, solar, and wind. As shown in the earlier table, the energy contributions from these are far less than the energy we use now. Without fossil fuels and nuclear, we could still have 17% of our present energy level. This new lifestyle is only possible if we increase our solar energy 120 times and our wind energy by 30 times in the next 50 years. We better get started.

- If there are others who have alternatives to The Five Percent Plan, they should quickly make themselves heard. The proposed plan (or any similar plan for that matter) is very time sensitive and needs to start immediately. If we don't get started, we will lose the time sliver of opportunity. We will have wasted the fossil fuels that our descendants will so desperately need over the next 150 years. By 2005 as this book is being republished, it may already be too late.

- Note again, as explained in Chapter 3, just the fuel that could be saved by minimal, legislated, automobile-fuel economy (27 mpg) versus light trucks and SUV's (18 mpg) is 0.44 EBBO per year. If we had moved the other way in the 1980's to more efficient vehicles available at the time (35 mpg), we would be saving almost one EBBO per year right now. Gas-guzzlers and frivolous transportation are directly robbing the future from my and your children. It would have been so easy.

The difference in fuel consumption, for the same distance traveled, between what we are now using with large cars and light trucks and what we would consume at 60 mpg, is about one-fourth all of the energy required for our future model in 2050. Are we the most intelligent species on the earth?

We need to start immediately to reduce our energy consumption by five percent per year. In many ways this action will improve our quality of life. It will also resolve related problems, such as climate change and pollution. The message as summarized in this book must reach a majority of the U.S. public and not stop with you as an individual reader. It's up to everybody to get involved. Civilization cannot survive unless individuals make it happen. Get involved. Pass it on. Please help make peak oil and fossil-energy depletion the most important issues in the world. The war on terror, world democracy, climate change, and global conflict are subserviant and related to the fundamental crisis of dwindling fossil energy.

Chapter **6**

Reductions of Energy Usage to Achieve the Five Percent Plan

It's time for a seventh-inning stretch. Since starting this project (first edition) last winter, the summer of 2003 arrived, the lawn is green, and another gallon of gas fueled the plowing of our one-fourth acre garden. Hopefully, the Iraq war is behind us (it wasn't by 2006) and was at least morally justified considering the outgoing regime's horrific human-rights record. From an energy perspective, however, some of the incredible amount of fuel used for the war and subsequent occupation will ultimately be replenished from the country being liberated. Roman Legions marched using grain (energy), minerals, and slaves taken from the countries they conquered. Sometime in the next few years there will no longer be enough fossil fuel to support our ability to impact the world, and geopolitics will take on new meaning.

Review of Several Contrasting Positions

The first three-day international workshop on oil depletion held in Uppsala, Sweden (May 23–25, 2002) was organized by Uppsala University and ASPO (the Association for the Study of Peak Oil and Gas). This meeting and a second-annual workshop held in

Paris in May of 2003 (and more recent meetings) are covered in great detail on **www.peakoil.net**. To quote the mission of the first meeting which was attended by experts from all over the world:

> *The Workshop has been organized to address the impact of resource constraints and natural depletion on the global supply of oil, one of Europe's most critical energy sources.*
>
> *The Workshop will examine geological mechanisms of oil generation, the processes of extraction, and the nature of depletion as imposed by reservoir physics.*
>
> *The Workshop will address also the unreliable nature of public data on oil reserves, the record of past oil forecasts, and the misunderstanding of the 'Limits of Growth' message of the 1970's. It will question the ability of conventional economics to handle a rapidly depleting resource, and draw attention to the impact of oil's decline on Europe's economic situation. It will ask why governments are not better informed, and why they fail to make appropriate preparation to ease the impact of the transition from growing to declining oil supply.*

After digesting this report and other considerable input from the web and myriad other sources, I feel even more concerned for the future of humankind. Many put a happy face on the coming crisis by pointing out progress in renewable energy but closer analysis of quantitative details reveals work done so far is absolutely miniscule considering the pace we must attain to reach energy sustainability. For instance, Janet Sawin has written a very informative chapter in *The State of the World 2003*[1] detailing the early solar and wind progress made in Europe and Japan. She also points out the U.S. lags far behind Europe. Like many authors, she does not quantitatively emphasize how insignificant renewable energy progress has been so far compared to the awesome quantity of fossil fuel we now consume and how quickly it is being depleted. When we run out of gas, it's not going to help to find two people instead of one to push the car, even though we've doubled that tiny energy effort. We would need 50 people working in relays to push the car any appreciable distance.

Next, I would like to compare two contrasting approaches to the subject of energy in the future. First, from a presentation to the British House of Commons by Colin J. Campbell, July 7, 1999[2]:

> *The title of my talk is "The Imminent Peak of World Oil Production." I would like to provide the evidence. It is, of course, a very large subject. There are colossal economic and political consequences. Indeed the very future of our subspecies - Hydrocarbon Man - is at stake.*
>
> *My qualifications for taking up your time are that I have spent the last 45 years studying the subject both directly and indirectly. I have evaluated hundreds of oil projects around the world. I have drilled many dry holes and even made a few discoveries. I have observed the oil industry from many angles, including its senior management. I have published two books and several papers on oil depletion.*
>
> *I think I can summarize the position into a few key points.... We now have a comprehensive understanding of petroleum systems.*
>
> - *It has become relatively easy to identify and map them, once the critical information has been gathered from seismic surveys and excavation bore holes.*
> - *The prolific generation of oil and gas was a very rare event in both time and place in the geological past.*
> - *The world has now been very thoroughly explored with the benefits of this new understanding and the high-resolution seismic surveys. About 90% of the world's oil endowment lies in just 30 major petroleum systems.*
>
> *Looking at the world as a whole, we see this growing deficit, discovery peaked in the 1960's with a 60 Gb surplus, but that has given way to a deficit of almost 20 Gb. We now find one barrel for every four we consume.*

(Note: My comments, remember this was written in 1999. A Gb is a giga barrel, which is one billion barrels.)

The general situation seems so obvious. Surely everyone can see it staring them in the face. How can any thinking person not be aware if it? How can governments be oblivious of the realities of discovery and their implications? How is it possible given the critical importance of oil to our entire economy?

It is fashionable to accuse oil companies of conspiracy, but I think the deception has more to do with differing mindset and objectives than conspiracy. Oil companies are in business to make money, not plan the world's future.

I should say something about getting more oil out of the reservoir. You will hear a great deal about that. Now that falling discovery is widely recognized, hopes are pinned on getting more out of what has already been found. I am sorry to say that reports of improved recovery are largely an illusion reflecting the reporting procedure rather than any particular technological improvement.

The bottom line is that there is a rounded one trillion barrels left to produce.

The U.S.–48 is a good example. Discovery peaked in the 1930's and production in 1971, despite ample technology, money, and incentive. In the North Sea, peak discovery around 1980 is now being followed by peak production. It should surprise no one.

One indisputable fact stands out. Discovery peaked 30 years ago. It takes no feat of intellect to conclude that we now face the corresponding peak of production.

Compare the above reality and pessimism with the following excerpts from a talk given by U.S. Energy Secretary Spencer Abraham to the Detroit Economic Club, February 7, 2003[3]:

Back on January 2, 2001, when President Bush asked me to serve as secretary, I recognized as we all do, the importance of energy to our economy and to our way of life, I have to confess, however, that I didn't fully appreciate then just how much my job would relate to this state and to our industries. But the facts are clear - more than two-thirds of the 20 million barrels of oil Americans use each day is used for transportation. As a result, a key part of my job is working closely with Detroit and the automotive industry to devise energy solutions for our nation's future.

Last May here at The Club, I noted that the 21st century will bring a huge increase in energy demand. Our analysts foresee a 54 percent increase for electricity ... 54 percent for natural gas ... and 47 percent for oil ... by the year 2025. At the same time we can expect to see similar or even greater increases in energy demand all over the globe.

It is difficult to believe that Campbell and Abraham are talking about the same planet.

Abraham goes on to say about President Bush's 1.7 billion dollar hydrogen initiative:

> *We believe that the hydrogen economy is our future, so today, the questions we face are how fast this effort should proceed, and whether it will be led by America or by others.*
>
> - *First, we must lower the cost of fuel cells by another factor of ten.*
> - *Second, we must lower the cost of hydrogen ... virtually all hydrogen manufactured today is produced from natural gas. If successful, fusion could well be the most cost effective source of hydrogen we will ever find.*
> - *Third, we must devise new methods to store sufficient amounts of hydrogen fuel aboard a vehicle to provide the consumer with sufficient range.*
> - *Fourth, and most critically, we will work to solve the overarching infrastructure challenges I just outlined.*
>
> *As a result of this effort, we estimate that by 2020, we could achieve mass-market penetration of hydrogen fuel cell vehicles and the availability of fueling stations nationwide.*

The above sounds like five miracles end to end, all by 2020 and nowhere in the EIA's forecast to 2025 is there provision for hydrogen. As a lifelong car enthusiast, mechanical engineer, and career product-development manager, I would not put all my 1.7 billion dollars into this basket. All the above research has been going on for years with little to show for it except research grants and fuel for our space program.

Science and math predict that the age of high-speed travel will end soon. The required concentrated energy will become increasingly scarce and there are far more important uses for what is left.

Referring to and re-emphasizing the U.S. energy consumption curve, past and future, shown in Chapter 3, note well that the first half of the bell-shaped probability curve took place in one long, eighty-year lifetime.

On this same bell-shaped curve, the second half of total recoverable (assuming it is all recoverable) crude oil also occurs in one lifetime of children being born today. Two lifetimes, that's all, for the energy party. This is a blink of a blink, in the human timeline of world history.

The actions we take today will directly affect our children's survival. We must act now or lose the small window of opportunity to climb down a ladder from this peak we're on. The alternative is to decline into a nation and world of chaos and misery.

If we really love America and our descendents, we must raise a common voice to demand sweeping legislation so all citizens (not just a conscientious few) begin immediately to reduce energy consumption.

THE FIVE PERCENT PLAN
(See Appendix 4 for Summary)

Following are the four major sectors of U.S. energy consumption where we must begin immediately to save energy. The first reductions are easy since we have the highest per capita consumption in the world. We'll be better off as a society and see corresponding improvements in overall health and environment.

1. Transportation
(4.25 EBBO, 28% of total energy consumption)

The age of fast long distance travel will end with the demise of fossil energy.

• Big Cars and Trucks to Smaller Vehicles

The first step (and easiest) of our controlled descent down the Five Percent Plan requires getting out of our gas-guzzlers. As pointed out in Chapter 3, the use of light trucks (SUV's) at 18 mpg compared to ordinary cars at 28 mpg is costing us an excess of 0.44 EBBO per year. By driving readily available 35-mpg cars, we could save an additional 0.42 EBBO per year for a total of 0.86 EBBO. These two steps alone would take care of our total first year's goal of the Five Percent Plan. I believe a majority of citizens are ready for and believe in this move but are reluctant to do so because of intimidation by oversized vehicles and present low-fuel costs. A combination of fuel taxes and rationing would encourage equitable restraint by everyone. If someone were in your garden stealing your food, you'd complain. Someone is stealing your and your children's energy. Raise a voice. Switching from big vehicles to little ones would reduce consumption, improve safety, and not hinder our lifestyle. We need legislation which encourages a profound change in our thinking. Are you part of the problem or the solution?

• Drive Less

The next easy step is to consider if each trip is really necessary. I remember World War II when my parents had gas stamps for 5 gallons per week. That was a time of war and an accepted temporary crisis. What could be more of a permanent crisis than the one we are facing in the next 5 to 10 years? Walking and bicycling instead of driving would make us healthier. Planning multiple trips into one, using public transportation, car-pooling, vacationing closer to home, all could easily be done. By reducing our travel by 50% (and using 35-mpg cars) we could take another giant step towards reducing energy consumption. In the year 2000, 280 million Americans traveled 2,525,000 million (over two trillion) miles in their cars and light trucks (SUV's) using almost 400-million gallons of gasoline each day. By reducing this excessive travel by one-half with 35-mpg vehicles, we would save another 0.86 EBBO.

This brings us down the ladder to the year 2006 on our controlled five percent bridge. This is easy. We've already saved (0.86 + 0.86) = 1.72 EBBO (11% of 2001 total energy), and we're safer and healthier for it.

We're starting to look like Europeans, little cars, short trips. People around the world are starting to respect us.

The economy has been very busy using still available fossil energy to melt down big cars to build smaller ones and, if priced right, automobile manufacturers could still make a profit. The little cars will keep us going in great style providing we don't increase the population and negate the gains we've made. Little cars will also make it safer for electric cars and hybrids to appear on the road. Obviously this massive change of our personal transportation habit will not happen in several years. It would be spread over the next 20 years in parallel with all the other energy reductions.

• **Alternative Vehicles**

Switching to even more efficient vehicles such as hybrids, small diesels (already popular in Europe today), and electric cars could easily increase our average gas mileage to about 60 mpg. This is well within the reach of today's technology. Who knows, maybe a little hydrogen derived from fossil fuels might be introduced, although the natural gas feed-stocks, complexity, required infrastructure, and drastically reduced emissions we've already achieved make hydrogen-powered vehicles superfluous and very wasteful. By going from 35 mpg to 60 mpg and traveling 50% less in our personal cars, we save an additional:

0.86 EBBO used at 35 mpg minus 0.505 EBBO used at 60 mpg = 0.36 EBBO

Our total energy savings is now (0.86 + 0.86 + 0.36) = 2.08 EBBO, or about one-half of our 2001 level of the total energy (4.25 EBBO per EIA) used for transportation. We would then be close to per capita gasoline consumption of modern Europeans. Obviously we need to use more public transportation (like Europeans) for longer trips.

• **Diesel Fuel**

In 2001, we used 0.934 EBBO for diesel fuel, about one-third the energy used for cars and light trucks. We can cut this in half just by slowing down and not shipping stuff from coast to coast. This would give us another 0.47 EBBO. Less interstate transport would encourage a renewal of local commerce and agriculture with a commensurate return to local employment.

I can hear the screams of anguish. "We can't shut down the economy this way." Listen to the laws of science and do the math. If we don't slow down on our terms now, in the next ten years civilization will begin to collapse, and we will spiral out of control with very little fuel left and no hope for correction. Mother Nature doesn't care about our future.

• Air Travel

In 2001, jet fuel amounted to 0.6 EBBO. We should start right away to also cut this in half giving us a total for both diesel and jet fuel of 0.77 EBBO.

0.47 EBBO diesel + 0.30 EBBO air = 0.770 EBBO

This brings us to a total energy consumption of 13.15 EBBO per year, the point we should reach in about four years after starting in 2005.

• Marine

Of course marine travel, like all other forms of moving people or things over a distance, will also have to decline and use wind and solar energy. Since this mode of travel is in the commercial sector, the quantitative reductions of energy will have to come from there.

• Military and National Security

It's not clear where U.S. government and military energy consumption is shown in the EIA summary. One reference reports 993×10^{12} BTU/year (0.158 EBBO) for the war year 1991.[4] Like all other sectors of consumption, military use must also be significantly reduced in the future. As a functioning nation, our primary concern, other than food and energy, is national security. For this reason, it is critical that we preserve our fossil-fuel resources as much as possible. I don't believe even the most optimistic dreamers can envision solar-powered jet planes or hydrogen-powered tanks. As the second half of our 200 year fossil-fuel era winds down, nations with fossil energy will be the ones that control their security. This is an extremely serious consideration which is underlying much of the world's geopolitical unrest.

Do our long-range military planners understand where the fuel for national security is going to come from 50 to 100 years from now? This is another reason it is especially important that we start immediately on our Five Percent per year reduction plan.

Our total U.S. petroleum reserves of 0.7 EBBO in the Louisiana Salt Mines and an estimated 7.0 EBBO (one year's consumption at our present rate) yet untapped resources in the Artic National Wildlife Reserve in Alaska (ANWR) will be far more important for our future survival than for frivolous use today or to sell on the world's oil markets.

• Recreation

All unnecessary use of fuel will have to end. A joy ride we enjoy today is future energy lost forever for humanity.

Now things are really getting serious. Some favorites that quickly come to mind are:

Internal combustion racing of any kind

A quick calculation of a single major Nascar race with 50,000 spectators driving, four in a car, 100 miles at 20 mpg plus transporters and 50 race cars shows a fuel consumption of 75,000 gallons of fuel. We should watch bicycle racing instead as the race comes through the villages to the spectators. In the three-week Tour de France, close to one-fourth of the population will see the race live, many without having to travel by car.

All other spectator functions

About 95% of the energy cost of automobile racing and other sporting events, is the fuel consumed by spectators. Similar reductions will have to be made in all activities where the mode of travel is other than foot, bicycle, or mass transit. How can we enjoy consuming fuel for pleasure when we will need it for future survival? Think about it. We should stay home and watch spectator sports on TV or participate in activities which emphasize human energy.

Other Sports

There are other very energy intensive sports; for instance, (one of my favorites) alpine skiing which combines personal

109

travel, second homes, or overnight facilities (heated and available), and extensive electricity for snowmaking and lifts.

School Buses and Soccer Moms

We probably use more energy for school transportation, sports, and related activities than many Third World nations use for everything. School travel will have to rely on public mass transportation, bicycles, or walking. I'm not dreaming this up. We must use less fuel. Our kids must learn what the low-energy future holds for them.

RV's, ATV's, Personal Watercraft, etc.

In light of dwindling and finite fossil-fuel resources, all of these activities will have to cease. This is not a personal vendetta. The party's over. (This was written before the book of the exact title appeared on the market.[5]) How can so few take from so many? These activities will have to be curtailed by legislation resulting in a combination of taxation and rationing. All, rich and poor alike, must share the new low-energy lifestyle.

Snowmobiles are just as bad. A typical 100 mile per hour sled gets 10 mpg. New 4 cycle machines are better but should be used only for absolute necessities. Why not go cross-country skiing or snow shoeing instead? I'm certain there would be a concurrent improvement in overall health.

Postal Service, Catalogue mail order marketing, and other delivery services

There is presently tremendous duplication and energy waste in the use of transportation energy. We can certainly do without unsolicited (junk) mail. We can use email, the web, and telephone for most communication. One or two package deliveries per week with substantial fuel surcharges to discourage all but the most essential shipments should keep our economy going just fine. Again, appropriate legislation will be required to effect these changes.

Space Travel and NASA

As concluded before, only the pinnacle of the free-energy age can support the effort of space exploration, and we are now passing that peak.

Why do we need further space exploration when our survival is at stake? Shouldn't we be focusing research and government subsidies to develop a fossil-fuel-free future?

---

2. Residential
(3.21 EBBO, 20% of energy consumption)

Residential energy use is almost as much as transportation. Of 3.21 EBBO per year, about one-third is for heating and two-thirds is for electricity. Using well know methods of conservation, turning the thermostats down, and being frugal with electricity, we should reduce this consumption by 50% in the next 10 years for a savings of 1.6 EBBO.

At the same ratio as present for heating and electricity, this new figure would give 1.1 EBBO for electricity and 0.5 EBBO for heating. If we assume 40 million homes, this would translate to 575 equivalent gallons of oil per year (most would be natural gas plus some wood) for heating—and for electricity, 34,000 kwh per year, or 2812 kwh/month.

This seems high for electricity for 40 million homes, but includes electric heat and substantial air conditioning. We should reduce this by one-half anyway. The above figures, at 50% present consumption, are still considerably more than the 200 kwh/month for electricity and 318 equivalent gallons of oil (from thermal solar) planned for the fossil-fuel-free residence (PR) in Chapter 4. The home heating consumption of 4800 kwh/yr ultimately planned from passive solar is equivalent to 215 equivalent gallons of oil burned at 50% efficiency. This energy would have to be supplemented by wood, coal, or oil and natural gas conserved from the 200 EBBO with our five percent per year reduction plan.

The above residential-energy consumption includes hot water, which would be divided between the electrical and heating allotments depending on local price and availability.

The total residential saving at 50% of the 2001 consumption rate is 1.6 EBBO. Subtracting this from our reduced annual consumption of 13.2 EBBO (due to transportation savings) brings us down to 11.6 EBBO or about where we should be in 7 years (2010) on our five percent reduction plan.

Included in the above energy reduction in the residential area is any energy associated with extra living convenience. Only summertime alternative living arrangements are acceptable much like the summer retreat common in the 1800's. Today's larger homes will have to shrink into a warm living core area for the winter months.

To keep cool, we'll just have to rediscover the measures our great grandparents used. We may have to suffer a little and reschedule our day's and evening's activities. The people in the north will be better in the summer and the opposite is true for the southerners. In the winter we'll turn the thermostat down, wear more clothes, heat fewer rooms, be creative. Again, we'll probably be healthier for it.

Everywhere in the rest of the world are small local gardens instead of lawns or wasted space. The return to home food production (residential micro-farming) will make us appreciate how much we've come to depend on fossil fuels and energy-intensive agribusiness. Remember, each adult requires about 500 pounds of high-quality food (energy) per year. Our personal energy allowed us to procure that food input as hunter/gatherers and farmers as long as we did not deplete the resources. We are much healthier when we have a good personal balance between fuel intake and energy expense.

A major cropland in the USA with 32 million acres is the lawn. We will have to return this area to supplemental food production. The energy used to mow lawns would be far better used for growing food.

---

3. Commercial
(2.77 EBBO, 17% of Energy Consumption)

In 2001, commercial energy use was 0.67 EBBO for primary fossil fuel consumption and 2.1 EBBO for secondary electricity of which approximately 75% also comes from fossil fuels. The commercial sector is going to be more of a challenge for energy reduction because of its direct impact on jobs and economic activity. In the transportation and residential sectors, we only have to make personal changes in our lifestyle. Most energy reduction will not hurt as we take several steps backwards towards our great grandparent's quality of life. For our step by step decent down the Five Percent Plan and still keep our economy going, we will reduce the primary commercial fuel by only 25% and secondary commercial electricity by 50% (operating buildings less hours, turn down the heat, turn lights off, improve energy efficiency, use less space, etc.). This gives us savings of 1.2 EBBO (0.67 times 0.25 + 2.1 times 0.5 = 1.22 EBBO). Added to the previous savings this brings us down to 10.4 EBBO energy consumption per year. Now we are down to where we should be by the year 2012.

4. Industrial
(5.18 EBBO, 32% of Energy Consumption)

The final and largest sector of our energy-intensive economy is industry which uses 5.18 EBBO per year at the present rate, proportioned as 3.43 EBBO as fossil fuels and 1.75 EBBO as electric power with approximately 75% also coming from fossil fuels. By reducing this consumption sector also by 50%, we save another 2.6 EBBO, which brings us down to 7.8 EBBO annual energy consumption. This is where we should be in the year 2018 on our Five Percent Plan and is about one-half of our present (2003) energy consumption.

This leaves us another 22 years to eliminate another 2 EBBO and reach our final goal of 5.84 EBBO per year by the year 2040. This entire decrease in consumption is shown in the curve in Chapter 5. The actual numbers are summarized again in Appendix 4.

If we do not take control of our descent down the five percent per year energy plan, we will be nearly out of fuel by 2040. Energy costs will soon go out of control. World civilization as we know it will be doomed because we did not save enough fossil energy to reach a level of sustainability. We will never get another chance.

Both controlled and uncontrolled descents will incur significant fossil-fuel reductions each successive year. The only difference is the Five Percent Plan occurs about 20 years earlier and leaves a very reasonable quality of life for our descendents. Leaving this critical issue another 20 years to the free-market place and short-term interests will expand world conflict, chaos, and misery as competing people fight for dwindling resources. We will then be effectively out of oil and natural gas with only dirty coal left, and can only hope to sustain a 3 EBBO per year level achieved with renewables, hydro, and nuclear, all of which depend on some fossil fuels for mining and manufacturing. Without the 200 EBBO of conserved fossil fuels we saved by starting immediately, 20 years sooner, civilization, as we know it will cease. We'll be out of gas.

Do we have the willpower to start saving now so our children will have a future? The next few years will tell. The 2004 election year was critical for determining our direction yet we made no change. By 2005 when this book was rewritten and gasoline is $2.30 per gallon, our second-term president is starting to at least mention the need for renewables as well as the "C" word, conservation.

Planning for the future sounds obvious and simple, but it may not happen because of human denial, greed, and unwillingness of our leaders to comprehend and educate their constituents about the facts and hard choices. The impending crisis and demise of civilization will exceed all preceding wars and social upheaval in devastation, and we cannot win. Our only hope is for a majority of the voting public to urgently get behind this subject and encourage appropriate legislation which forces everyone to start now down the Five Percent Plan. We have no time to lose. Continuing to wobble along the top of Hubbert's Peak for the next few years will squander the most critical time at our present, prodigious, consumption rate.

If this scenario is overly pessimistic by a few years, hurray. It just means we have a little more time to get our act together and face the most important task ever to challenge humankind. By 2005 the energy situation definitely appears more pessimistic as energy (and all commodity) prices continue to increase.

Chapter 7

Related Issues, Summary, and Call to Action

GEOPOLITICS

We now have a quantitative understanding of the short, fossil-fuel Energy Age of which only a small minority of the world's population is enjoying. Day by day world news takes on new meaning. A large part of the international chaos and turmoil has a major energy (or lack of energy) component. The Industrialized World is intent on gaining a disproportionate share. Much of the Third World is already out of control because of over-population and dwindling or otherwise devastated natural resources made possible because of low-cost energy and transportation. The 2003 issue of *State of the World*[1] identifies many energy related problems such as forest and soil depletion, mining devastation, and squalid slums adjacent to and living off the dregs of energy waste from major cities. Another comprehensive reference, which explores the relationship between resources and population, is the book, *How Many People Can the Earth Support*.[2] It is especially discerning and prophetic to read the beginning of Chapter 16, "Case Study: Easter Island." The author quotes

117

from his subject reference, "We consider that Easter Island was a microcosm which provides a model for the whole planet."[3]

In this case, a prolific isolated island of 166-square kilometers was completely denuded of trees in about a millennium. The civilization, which prospered and slowly grew to about 10,000 people, totally collapsed. Left behind were huge statues, up to 65-feet long standing and scattered as a testimonial to the short-time energy surplus the natives enjoyed until increasing population and declining resources reached a collision point. It is interesting to note that the population crash occurred in only a couple of centuries compared to many more earlier years of good times. This completely isolated society lasted for many hundreds of years more than expected for our fossil-fuel Industrialized Age.

Four other notable books, which describe endless examples of resource devastation and population collapse, are *The Last Hours of Ancient Sunlight, Collapse, the Collapsing Bubble* and *A Green History of the World* all reviewed in my recent books list in Chapter 2.

While wandering around the web exploring references and links from **www.hubbertpeak.com**, I came across an interesting paper titled, *Busby Report, UK survival in the 21st Century*[4] and to quote the cover page:

> *The sun has set on the 20th century, when the "Black Gold" of oil provided the mobility for the creation of a global community. This report considers the measures to be taken to ensure the survival of the United Kingdom in a new century during which the world's oil will run out.*

Mr. Busby goes on to say,

> *We also have a huge task in preparing the British people for a change in lifestyle before a more unpleasant change is forced upon them by circumstances and to be able to re-direct capital and resources in a direction not currently taken, espe-*

cially as it may contrast with that taken by the rest of Europe or of the world.

As we face a future of global collapse in 25 to 50 years time, surely if the situation is properly explained and a sensible and considered plan for survival through the 21st century is promoted, the nation will rise to the task ... the time table has started.

Whether the system dynamics forecasts are accurate is a question merely of timing. If global economic collapse does not occur around 2050, it will before the century is out. So national planning for survival runs only the risk of being a little premature.

We may have to unite in a united states of Europe to withstand the economic depredations of a USA wanting to secure more than its share of dwindling energy and other resources. Progress in alternative energy is currently more advanced in Europe, in wind farms and with bio-diesel available at filling stations. But if we put our own renewable resources in place in the next crucial 10 years or so, while at the same time reorganize our economy to use a quarter of our current energy requirements, we will provide a model for a second industrial revolution in Europe and renew our world leadership in ideas.

Busby then goes on to offer a plan for national priorities in many areas and then offers two scenarios for Britain in 2025.

Here are two extreme scenarios—The first assumes that none of the recommended actions have been taken, the second assumes that all have been implemented.

The first scenario goes on to describe a Europe in chaos combined with U.S. military reaction to maintain its threatened lifestyle. The first scenario concludes with:

However communications developed so that it is quite normal to see a horse and cart with the carter discussing customer requirements with a mobile telephone.

119

The second scenario goes on to describe a very acceptable economy and lifestyle with better health due to:

> *Fewer accidents on empty roads, participation in healthy sports and local activities. Drug-taking was finally brought under control not by policing but by the reduction in international traffic.*
>
> *Unemployment is not the problem it might have foreseen to be as the lack of energy, to drive machinery, means that many functions have returned to manual labor.*

... and so, on and on. Here you have an indication of recognition of the immediate pending crisis combined with a semblance of a plan and hope for a very acceptable future. Busby's plan is based on working down to 25% of the present U.K. energy consumption whereas the plan in Chapter 5 in this book is based on a U.S. reduction to a sustainable level that is closer to 40% of present consumption. This should be a much simpler task especially considering that we in the U.S. presently waste considerably more energy than the people in U.K. This is in large part because of $1.50/gallon instead of $4.50/gallon gasoline (2003 figures).

The message throughout the world is, the U.S. is seen as the nation of bad-energy guys, and inevitable future turmoil will further identify and isolate U.S. as the gluttons of world energy. Geopolitically, we need to reckon with this honest perception and use it to explain much of the evening news.

Rather than encumbering this short book with the complex synergism between geopolitics and energy, I'll leave the topic to many, more qualified sources. Numerous authors cited below and in other chapters, devote considerable effort in the attempt to unravel the interaction of world politics and finite resources.[5,6,7,8,9] Obviously, these two subjects have been inextricably linked since the beginning of history. If we go back even farther, we can see the same phenomena taking place at all levels of biology and nature from squash bugs and potato beetles competing for plant space to mice overrunning a barn until depleted resources or cruel population-control limits their numbers.

RELIGION

In these troubled times, the Islamic Middle East, which sits on top of most of the remaining world oil resources, is especially volatile. Millions of the younger generations are rendered irrelevant as more advanced societies move in, obviously interested only in the energy of their black gold. (Sounds a little bit like Spaniards and Aztecs). These indigenous people (rich and poor alike) turn to their strong religious fundamentalism to find answers, which in extreme cases clearly leads to terrorism. This is their only possible response to a vastly superior military presence. It is argued that oil is not the issue, but it certainly appears to be an underlying concern and common thread.

On a positive note, there is an inherently strong common bond between all religions and the environment. *The State of the World 2003*[1] concludes with an excellent chapter by Gary Gardner entitled: "Engaging Religion in the Quest for a Sustainable World." Mr. Gardner concludes:

> *... by combining their considerable skills and complimentary perspectives, environmentalists and religious people can help re-unite our civilizations head and heart, re-engaging religion in the quest for a new cosmology, a new world view for our time.*

Another example is the conjunction of The National Council of Churches and the Sierra Club to fund expensive public ads to discourage drilling in ANWR.

Of 6 billion people in the world, Christianity (2 billion) and Islam (1.2 billion) are the largest. These religions, as well as the others like Hinduism (0.75 billion), Buddhism (0.36 billion), and Judaism (13 million) all sincerely teach respect and stewardship for the natural world.

All successful indigenous societies had to have a strong environmental connection to be compatible with the laws of sustainability. If they didn't, it's Easter Island all over again.

Long-term civilizations usually survived until climate change, invasion from outside, or resource depletion upset their delicate energy balance with nature. Keep in mind, the great historic civilizations, like Mesopotamia, Egypt, Greece, and Rome lasted many more centuries than our present two-century, fossil-energy age. It appears, if outside forces threaten a society, strong religious ties work together to resist and protect that society's resources. But then, religion may not confront the most obvious cause of resource depletion, excessive population, which flourishes on bountiful resources and good times. Perhaps the classic confrontation between good and evil can be interpreted as a delicate interplay between consumption and supply. The periodic plagues of locusts or natural drought can be an interpretation of God's wraths for various sins. Maybe the Egyptians were closer to the truth in focusing on the sun god, Isus, as most important to life on earth with the sun as the source of life-giving energy.

If God endowed us with a superior brain, he would expect us to use our skills to read the past, plan for the future, and do a far better job of resource utilization and conservation.

To quote the summarizing paragraph in the *State of the World 2003* reference,

> *By combining their considerable skills and complimentary perspectives, environmental and religious people can help reunite our civilizations head and heart, re-engaging religion in the quest for a new cosmology, a new world view for our time. Cultural historian, Thomas Berry, calls this emerging perspective a New Story—the story of a people in an intimate and caring relationship with their planet, with their cosmos, and with each other.... It would be the vehicle to guide us to a socially just and environmentally sustainable future.*[10]

Since writing the first edition of my book I've had the good fortune to meet Dr. John E. Carrol. His recent book *Sustainability and Spirituality*[11] directly addresses the confluence between these two "rivers of thought". The synthesis of religion and science leads to ecology, the study of our place in our earthly home.

If the human epoch is to survive we must find a spiritial connection and true sustainability with our natural environment.

I will conclude these thoughts about religion and sustainability by citing two other references. First, an interesting book came into my possession through a library book sale entitled *Cracking the Apocalyptic Code* by G. Bodson.[12] The authors describe the painstaking, numerical cracking of the apocalyptic code buried in the last book of the *Bible*, "The Book of Revelation," which "presages the end of humanity itself." This team of five experts worked laboriously together to expose the hidden prophecies and determine the date of Armageddon, the final battle between good and evil. The date they came up with is February 5, 2043 at midday. It's disturbing to note that as of 2003 most of the world oil experts are telling us we have one trillion barrels left from the total original endowment of two trillion barrels at the beginning of the Industrial Age. We are using it at a rate of more than 28.5 billion barrels per year. Is it coincidence that this 35-year remainder to production ratio approximates the prophecy?

Secondly, I would like to call attention again to an excellent feature in the Nov/Dec 2002 issue of *E-Magazine*.[13] The subheadings in this well researched article tell the story:

- "The Growing Religious Mission to Protect the Environment, Stewards of the Earth"
- "Science and Religion Meet"
- "Energy and Climate: Galvanizing Issues"
- "Active for Islam"
- "The Buddhist Way"
- "Hinduism's Green Tradition"
- "Jewish Reverence for Life"
- "Theological Dissent"

Only the last subheading, "Theological Dissent" offers contrasting opinions from the other side invoking "The Pastoral Division between God's Mission for Humankind: Stewardship or Dominion."

As with geopolitics, there is a strong connection between energy and religious interpretation for the rise and fall of mankind's fortunes. These interactions may not always be obvious, and sometimes they are contradictory. Using energy as a common theme may help clarify confusing and chronic questions all the way from the Bible to contemporary population issues.

SUMMARY AND CALL TO ACTION

So there you have it. An undeniable crisis is facing mankind. We can quibble about the details and exact timing, but there is no question that we've enjoyed a steep and pleasurable 100-year climb to a very precarious peak. The view is great, but we cannot linger a minute too long. We have to plan a careful descent while we still have fossil fuels left to support our five percent per year descent to a low-energy future.

Obviously, the greatest task is to reach and convince a majority of the voting American public that all is not well. This is extremely difficult considering the short-term happy face, feedback loop of a constituency, which elects representatives and leaders only to hear good news and avoid the bad. In an election campaign, bad news is a "no-no" except if the leadership can invent dragons, which make them look good while slaying them, an age-old ploy called "job justification."

The real dragon is the coming energy crisis, but no one wants to hear the gloom and doom Cassandras and the fossil-energy dragon appears too formidable to challenge.

We are like lemmings reputed to follow each other to their demise in the sea, a harsh form of population control. Short-term economic growth and prosperity takes precedence over long-term choices even if they are obviously contradictory. We will be no different than other animals if we don't consider the past, do the science and math, predict the future, and take appropriate, timely action. The timing and circumstances are critical and momentous. Consider the following 1989 quotation by William

D. Ruckelshaus, who was the administrator of the EPA under Nixon and Reagan.

Can we move nations and people in the direction of sustainability? Such a move would be a modification of society comparable in scale to only two other changes: The agricultural Revolution of the Late Neolithic and the Industrial Revolution of the past two centuries. These revolutions were gradual, spontaneous, and largely unconscious. This one will have to be a fully conscious operation, guided by the best foresight that science can provide.... If we actually do it, the undertaking will be absolutely unique in humanity's stay on the earth.[14]

If we collectively don't act soon, in the next few years, all is lost. A couple more years of our unchecked, gross energy consumption will preclude any quality of life for our children. Look around and look to ourselves. Who are the takers?

How do we make something happen? This is the most important part of the book.

- **Project the Message in any form, in any way.** The importance of diminishing fossil-fuel resources far exceeds any other subject except possibly personal health to which it is indirectly related. Energy is certainly a part of food, shelter, work, recreation, and travel. Every part of your life should be reconsidered in terms of energy consumption. Raise a common voice. If individuals don't make it happen, all is lost.

- **Start to monitor world news** especially as related to energy. Read through the smoke and mirrors and put all related pieces of information into your own big picture. Check at least some of the websites and books referenced in this book. Cross-link to other sites.

- **Reconsider comprehensive planning**, short-term and long-range. All future plans related to energy will have to change. This especially includes family planning. Can the world provide for new mouths to feed when we won't have enough to go around? Will the coming quality of life be acceptable for us and our descendents? Look no further than sub-Saharan Africa, Ecuador, or India. Some parts of China are enjoying a temporary reprieve because of the influx of fossil energy and consumerism. Rural areas, however, are struggling with subsistence survival combined with dwindling resources. Unfortunately, even immediate negative population growth will not quickly reduce the population component of the energy equation. It takes several generations for a decreased fertility ratio to take effect as the bulge of population numbers moves through the years towards old age. By buying time using the five percent per year reduction plan, we have the chance to eventually reduce population at a controlled rate proportional to decreased energy. Think of this fact—if (like energy consumption by 2150) we are able to reduce U.S. population to about one-fourth of the present 280 million, the per capita energy would be the same as today. Even if we reduce our U.S. population by one-half with one-forth of today's energy consumption, we could have the same high-energy lifestyle as other industrialized countries enjoy today. By then, in 2150, hopefully we'll have worked out how to combine new technology with low-energy lifestyles.

- **All community or national capital projects should be reconsidered**. Does it make sense to add a third lane to an interstate or expand an airport facility? We should be downsizing and questioning all planning, personal and public, in terms of future energy availability. We need to plan on a future with drastically reduced travel, heating, and air conditioning to name just a few of the many forms of plentiful energy to which we have become accustomed. Immediate investment should begin to quickly expand the manufacture of photovoltaic cell and wind generating equipment. Plans should be made to recycle large vehicles into 60-mpg diesel or hybrid cars. These will be the last generation of personal vehicles before fully electric cars become the only alternative to the bicycle or walking. The work force should be directed towards a low-energy future instead of perpetuating the past. Millions of new jobs will be created for the transition and new low-energy lifestyle.

As the first edition of this book was being finalized, the Ford Motor Company celebrated its 100th anniversary. (There's that 100-year time span again.) So what are they doing? They're introducing a new, bigger and better, F150 pickup truck for the public, and at the same time planting grass on their new factory roofs to appease the environmentalists. By 2005 Ford's (and General Motors') fortunes are in a nosedive and their stocks have achieved junk-bond status. The age of SUVs and pickup trucks for personal transportation is over.

- **Personal Changes**. Make plans now to move towards energy efficiency, reduced travel, smaller insulated homes, solar and wind generation, growing your own food, and reduced consumerism. This especially should exclude the marketing and advertising of energy-intensive products. Of course, there'll be immediate changes required by everyone, but it is our only chance. Don't believe that things are happening automatically. Inertia continues on the easiest path. You've heard much of this before, yet we're still moving steadily in the wrong direction. A few years ago when oil was cheap, talk of an energy tax was dismissed as devastating to our economy. Now we're paying a huge "energy tax" to oil companies and oil producers in the form of higher prices.

We must gain a majority in our democracy so the required action becomes law for all. The future lies with with legislation (rationing, credits, and taxation) directed towards food, shelter, warmth, reduced population, and any other activities that use only a minimum of energy.

127

THIS BOOK

The purpose of writing this book is to reach the public on the most important subject of fossil-energy depletion. Hopefully I've made the reader aware of the comprehensive story of energy and fossil fuels. The expense for publishing is coming from my pockets and children's inheritance for obvious reasons—it is their only chance for a future. This book is copyrighted and may be reproduced (except for resale or profit) and with proper credit. The intent is for anyone to use the message, as is, or in any way or form to reach exponential numbers of readers as quickly as possible. Any number of additional complete books can be purchased individually for $10.00 each, which covers further printing, handling, and shipping costs. Volume pricing is approximately fifty percent depending on quantity. For further information write to Howe Engineering Company, 298 McIntire Road, Waterford, ME 04088, or contact McIntire Publishing Services at **www.mcintirepublishing.com**.

There is no glory being the harbinger of bad news. There is nothing new or unique in this book. It is only a compilation of existing facts, science, math, and logic.

All I've attempted to do is to distill and simplify the huge subject of energy, especially in the finite fossil-fuel form, and how it will affect civilization as we know it. I've tried to offer an engineer's vision of a downstream, low-energy, but still industrialized lifestyle. My career work has been in the General Electric Co. and AMF/Head Ski Co. I have manufactured and marketed bicycle-powered generators and grain threshers. It was this work, that first convinced me how feeble the human body is for converting food energy to useful work. I will continue, in every way possible, to further study, monitor, and project the message of declining fossil fuels. I have also constructed a prototype of a combination electric-transportation vehicle and small farm tractor or SPUV (see Chapter 4, the back cover and Appendix 7). A machine like this may realistically be the only hope for an industrialized post-fossil-fuel future.

My SPUV is powered from an on board photovoltaic system. I hope to prove (as many others have already) that a reduced energy lifestyle will be very acceptable and more rewarding than the forced consumerism treadmill that now controls us. By summer of 2006 I hope to have on the road a solar-powered 1962 MG Midget.

The future of the civilized world (hopefully led by the U.S.) depends on acceptance of and action taken from this and many similar messages.

Don't be deluded into thinking that the problem is non-existent or is being overstated. The problem is our gross consumption of energy. We are the problem, and if we don't change there is no hope.

In conclusion, I offer a quotation from a lecture series titled "Of Men and Galaxies" given 39 years ago in 1964 at the University of Washington by cosmologist Sir Fred Hoyle:

It is often said that, if the human species fails to make a go of it here on earth, some other species will take over the running. In the sense of developing high intelligence this is not correct. We have, or soon will have, exhausted the necessary physical prerequisites so far as this planet is concerned. With coal gone, oil gone, high-grade metallic ores gone, no species however competent can make the long climb from primitive conditions to high-level technology. This is a one shot affair. If we fail, this planetary system fails so far as intelligence is concerned. The same will be true of other planetary systems. On each of them there will be one chance, and one chance only.

This is exactly the prediction proposed in the *Olduvai Theory* discussed at the end of Chapter 4.[15]

To repeat an earlier statement, civilization will not survive unless individuals make it happen. Get involved. Pass it on. It is especially important, considering that time is of the essence, that we reach influential leaders or media that can, in turn, quickly reach a much broader audience. We cannot give up, no matter how futile our efforts may at first appear. Your initial depression will be followed by denial and finally a happier feeling as you become proactive by atempting to save humankind from a terminal illness. Continued, uncontrolled population and growth on a finite living biosphere (the earth) has been likened to a cancerous tumor. We must cut off its like support (fossil energy and other dwindling resources) and undergo whatever radical treatments necessary to get it under control.

endnotes

Chapter 1

[1] Seidel, P., *Invisible Walls: Why We Ignore the Damage We Inflict on the Planet ... and Ourselves*, (New York: Prometheus Books, 1998).

[2] Deffeyes, K., *Hubbert's Peak: The Impending World Oil Shortage*, (New Jersey: Princeton University Press, 2001).

Chapter 2

[1] Smil, V., *Energies*, (Massachusetts: MIT Press, 1999).

[2] www.world-nuclear.org

[2] www.eia.doe.gov/energyupdate2005

[3] www.asponews.org/nosi

[3] Rifkin, J., *The Hydrogen Economy*, (New York: Parcher/Putnam, 2002).

[4] Hoffman, Peter, *Tomorrow's Energy*, (Massachusetts: MIT Press, 2002).

[5] Lovins, Amory, various papers on hydrogen energy available from R.M.I. (The Rocky Mountain Institute).

[6] Bockris, J., *Energy, The Solar-Hydrogen Alternative*, (New York: John Wiley & Sons, 1975).

[7] www.hydrogenus.com

[8] www.hydrogennow.org

[9] www.fromthewilderness.com (see articles on hydrogen as well as many more referring to peak oil).

Chapter 3

[1] Deffeyes, K., *Hubbert's Peak: The Impending World Oil Shortage* (New Jersey: Princeton University Press, 2001).

[2] Campbell, C.J., *The Essence of Oil &Gas Depletion* (Multi-Science Publishing Company and Petroconsultants, 2003).

[3] www.eia.doe.gov, (Energy Information Administration), 2001, 2002, 2003.

[4] Stobaugh, R., and Yergin, D., Editors, *Energy Future: A Report of the Energy Project at the Harvard Business School* (New York: Ballantine Books, 1979).

[5] Yergin, D., *The Prize* (New York: Free Press, 1991).

[6] Meadows, D.H. and D.L., *Beyond the Limits, Confronting Global Collapse, Envisioning a Sustainable Future* (Vermont: Chelsea Green Publishing, 1992).

[7] Passerini, E., *The Curve of the Future* (Alabama: The Environmental Action Clearing House, 1991).

[8] Rifkin, J., *The Hydrogen Economy* (New York: Tarcher/Putnam, 2002), Chapter 2.

[9] Behr, P., "Of Drills and Energy Bills," *The Washington Post*, 26 May - 1 June 2003, p. 20.

[10] Bradsher, K., *High and Mighty* (New York: Public Affairs, 2002).
[11] www.bts.gov (Bureau of Transportation Statistics).
[12] Gore, A., *The Earth in Balance* (New York: Houghton Mifflin, 1992), p. 173.
[13] www.worldwatch.com
[14] www.cees.com, or www.bu.edu/cees
[15] www.nrel.gov

Chapter 4

[1] Scheer, Dr. H., *The Solar Economy* (London: Earthscan Publications, 2002).
[2] Pimental, David and Marcia editors, *Food Energy and Society* (University Press of Colorado, 1996).
[3] Duncan, R.C., *The Olduvai Theory* (Institute on Energy and Man, 1996) www.greatchange.org/othervoices.
[4] Pfeiffer, Dale Allen, *From The Wilderness* newsletters. Oct., Nov., Dec. 2003 www.fromthewilderness.com.

Chapter 5

[1] Klare, Michael, *Resource Wars: The New Landscape of Global Conflict* (New York: Henry Holt and Co., 2001).
[2] Kaplan. Robert D., *The Coming Anarchy: Shattering the Dreams of the Post Cold War* (New York: Random House, 2000).
[3] Homer-Dixen, Thomas, *Environment, Scarcity, and Violence* (New Jersey: Princeton University Press, 1999).
[4] Duncan, R.C., *The Olduvai Theory* (Institute on Energy and Man, 1996). www.greatchange.org/othervoices.
[5] Price, David R., www.greatchange.org
[6] Heinberg, R.,*The Party's Over: Oil, War, and the Fate of Industrialized Societies* (B.C. Canada: New Society Publishers, 2003). www.newsociety.com
[7] Thompson, B.
[8] Korpela, S., www.greatchange.org/othervoices.
[9] Grover, J. et al., *Beyond Oil: the Threat to Food and Fuel in the Coming Decades* (Colorado: University Press of Colorado, 1991).
[10] www.apolloalliance.org

Chapter 6

[1] The World Watch Institute, *State of the World 2003* (New York: W.W. Norton, 2003).
[2] Campbell, C.J., "The Imminent Peak of World Oil Production," presentation to the House of Commons, 7 July 1999.
[3] Abraham, Spencer (U.S. Energy Secretary) remarks to the Detroit Economic Club, February 7, 2003, www.eere.energy.gov.
[4] EIA, *Energy Review* (1991 to 2001), fig.1.3.
[5] Heinberg, Richard, *The Party's Over: Oil, War, and the Fate of Industrial Societies* (B.C. Canada: New Society Publishers, 2003).

Chapter 7

[1] World Watch Institute, *State of the World 2003* (W.W. Norton), www.worldwatch.com.

[2] Cohen, Joel E., *How Many People Can the Earth Support* (W.W. Norton, 1995).

[3] Bahn, Paul G., and Flenley, John, *Easter Island, Earth Island* (New York: Thames and Hudson, 1992).

[4] www.afteroil.co.uk

[5] Heinberg, Richard. *The Party's Over: Oil, War, and the Fate of Industrial Societies* (B.C. Canada: New Society Publications, 2003), www.newsociety.com.

[6] Yergin, D., *The Prize, The Epic Quest for Oil, Money, and Power* (New York: Free Press, 1991).

[7] Rifkin, J. *The Hydrogen Economy* (New York: Tarcher Putnam, 2002).

[8] Meadows, Dennis and Donella Meadows, *Beyond the Limits* (Vermont: Chelsea Green Publishing, 1992).

[9] Brown, A., *Oil, God and Gold: The Story of Aramco and Saudi Kings* (New York: Houghton Mifflin, 1999).

[10] Swimme, T., and Berry, *The Universe Story* (New York: Harper Collins, 1992).

[11] Carroll, John E., *Sustainability and Spirituality* (State University of New York Press, 2004)

[12] Bodson, G., *Cracking the Apocalyptic Code* (Boston: Element Books, 2000).

[13] *E-Magazine*, Volume XIII, No. 6, Nov./Dec. 2002.

[14] Ruckelshaus, W.D., *Scientific American*, "Toward a Sustainable World," Sept. 1989.

[15] www.greatchange.org/othervoices

Appendix 1

Solar-Energy Calculation

40 million homes each with a 2 to 4 peak kilowatt photovoltaic system (depending on location) could provide an average of 400 kwh/m or 4800 kwh/yr. (6.6 hrs. to 3.3 hrs./day average sunlight):

$$40 \times 10^6 \times 4800 \text{ kwh/yr/home} \times 3{,}412 \text{ BTU/kwh}$$
$$= 655 \times 10^{12} \text{ BTU}$$

$$655 \times 10^{12} \text{ BTU} \times 0.159 \text{ EBBO}/10^{15} \text{ BTU}/.35 \text{ efficiency}$$
$$= 0.3 \text{ EBBO}$$

(This compares to the 2001 solar U.S. photovoltaic production of 0.01 EBBO.)

Add residential non-photovoltaic (passive and active) solar for hot water and domestic heating equal to the photovoltaic component:

$$0.3 \text{ EBBO} + 0.3 \text{ EBBO} = 0.6 \text{ EBBO}$$

Photovoltaic output from centralized (CDF) power plants would equal the total solar output from residential (PR) distributed sources. The total solar output (photovoltaic and non-photovoltaic) from all sources would be:

$$0.6 \text{ EBBO} + 0.6 \text{ EBBO} = 1.2 \text{ EBBO}$$

In other words, it would take 1.2 EBBO of fossil-fuel input from power plants to provide the direct solar-energy output expected from our model.

To summarize the proposed annual U.S. photovoltaic-electricity production by 2050:

$$0.3 \text{ EBBO (PR)} + 0.6 \text{ EBBO (CDF)} = 0.9 \text{ EBBO}$$

This is the equivalent fossil fuel that would be required at 35% efficiency to produce the end use (secondary) annual electricity as follows:

$$192 \times 10^9 \text{ kwh (PR)} + 384 \times 10^9 \text{ kwh (CDF)}$$
$$= 576,000 \times 10^6 \text{ kwh}$$

It will be an incredibly daunting task to reach this solar-photovoltaic output by 2050. At an average insolation of 150 hours/month or 1800 hours/year, the total installed photovoltaic production capacity in U.S. alone would have to be:

$$576,000 \times 10^6 \text{ kwh/yr. divided by 1800 hrs./yr.}$$
$$= 320,000 \text{ Mw}$$

For the next 45 years until 2050, this much capacity would require an average PV production and installation of 7,111 Mw each year. As of 2002, the PV cell/module production in U.S. was 120 Mw and in the total world it was 562 Mw.[1] This means we must increase U.S. photovoltaic production sixty times (!!) in order to move to a true solar-powered economy. Obviously, we need to get started as soon as possible. If we wait for the market forces and prices of declining fossil fuels to motivate us, it will be too late.

Footnotes:

[1] PV News, Vol. 22, No. 3, March 2003.

Appendix 2

Human-Energy Requirements

We will start with 2000 kilocalorie (C) average energy intake required daily by 200 million adults. This equates to an annual energy requirement as follows:

$$2000C \times 200 \times 10^6 \times 365 \times 3.968 \text{ BTU/C}$$
$$\times 0.159 \text{ EBBO}/10^{15} \text{ BTU} = 0.092 \text{ EBBO}$$

If we add another 90 million children each requiring 1,000 kilocalories per day, we find a total U.S. energy intake for food of 0.113 EBBO.

If we assume an approximate factor of 10 to find the energy required to bring this food to our table, we end up with about 1 EBBO per year just to feed our present population. This tenfold increase includes the energy for crop tillage, planting, fertilizer, harvesting, processing, waste, packaging, presentation, and marketing.

There is much confusion and complexity regarding this subject; for instance, it has been estimated that the energy content required to feed animals to be subsequently used for human consumption is ten times the energy used if the plant food went directly to human use. The intent in this (Appendix 2) discussion is only to show the order of energy magnitude just to fuel our most important machines, our bodies.

Appendix 3

Planned 50% Reduction in Residential-Energy Use

We will start with the EIA total for the residential sector in 2001 of 20.2 x 10^{15} BTU or 3.21 EBBO. Of this total 6.9 x 10^{15} BTU or 1.1 EBBO is primary consumption of fuel for heating and the balance of 2.1 EBBO is secondary energy used as electricity.

The heating component of 1.1 EBBO is equivalent to:

$$\text{1.1 EBBO x 42 gal/barrel} = \text{46.2 x } 10^9 \text{ gal}$$

If we assume 40 million homes, this works out to 1,150 gallon oil (or equivalent natural gas) per home. Multiplying this by one-half means we will still have 575 gallons per year for heating. Anywhere else in the world, this much energy for heating would be positively opulent. With the proposed five percent reduction plan, even this home heating energy will have to be further reduced to eventually approach only what is available from non-photovoltaic solar. In Chapter 4, a typical solar home is described with 400 kwh per month of electricity component and an equal amount of solar heating. On a yearly basis the non-photovoltaic would be:

$$\text{4800 kwh/yr. x 3412 BTU/kwh x 1 gal./52,000 BTU}$$

(Equivalent oil is assumed at 35% efficiency or 0.35 times 150,000 BTU/gal.)

138

This equals 314 gallons per home with no fossil fuels used down to about 55% of our proposed interim level of 575 gallons per year.

The purpose of this exercise is to show the declining levels of home-heating energy (including some hot water) we will have to go through to reach a sustainable civilization.

The electrical component of 2.1 EBBO is equivalent to:

$$2.1 \text{ EBBO}/0.159 = 0.35 \text{ EFF. x } 13.21 \text{ x } 10^{15} \text{ BTU}/ 3,412 \text{ BTU/kwh}$$

This equates to 1.35×10^{12} kwh. If this much energy is provided to 40 million homes it would equal 2,812 kwh/month. This seems high for residential usage but includes considerable electric heating and air conditioning.

If we reduce this by 50% per our Five Percent Plan it still leaves us with 1,406 kwh/month. Far more than the 400 kwh/month estimated for both residential electricity and personal transportation in Chapter 4. There is considerable room here to allow for additional reductions in residential-electricity consumption and reach our goal of 5.85 EBBO by the year 2050.

Appendix 4

The Five Percent Plan:

A TIME LINE FOR A CONTROLLED DESCENT AND SYSTEMATIC ENERGY REDUCTION TO REACH A LOW-ENERGY FUTURE

The following summary table shows the total U.S. energy consumption goal for each successive year at a five percent annual reduction. These goals are compared to actual reduction that could be achieved as proposed in Chapter 6. Obviously, by 2006 and publication of the third edition, we already have lost several years and must shift the subsequent years accordingly.

Year	Five Percent Goal	Actual Reductions Achieved Per Chapter 6
2004	16	Start Five Percent Plan on Jan 1, 2005 at 2003 U.S. consumption level of 16 EBBO
2005	15.2	Large cars and trucks @ 18 mpg to 35 mpg cars. Annual savings = 0.86 EBBO. New level = 15.14 EBBO
2006	14.4	Reduce personal travel by 50%. Annual savings = 0.86 EBBO New level = 14.28 EBBO

(continued on next page)

Year	Five Percent Goal	Actual Reductions Achieved Per Chapter 6
2007	13.72	Switch from 35 mpg to 60 mpg vehicles. Annual savings = 0.36 EBBO New level = 13.92 EBBO
2008	13.03	Curtail diesel fuel and airplane travel by 50%. Annual savings = 0.77 EBBO New level = 13.15 EBBO
2009 2010	12.38 11.76	Reduce residential energy consumption by 50%. Annual savings in electricity = 1.1 EBBO. Annual savings in primary fuel = 0.5 EBBO. Total annual savings = 1.6 EBBO. New level = 11.55 EBBO
2011 2012	11.17 10.61	Reduce commercial consumption by 25% of primary fuel and 50% of electricity. Total annual savings = 1.2 EBBO. New level = 10.35 EBBO
2013 2014 2015 2016 2017 2018	10.08 9.58 9.10 8.65 8.21 7.80	Reduce industrial consumption by 50%. Annual savings = 2.6 EBBO. New level = 7.75 EBBO or where we should be by 2018. This gives us another 22 years to reach our sustainable level goal of 5.85 by the year 2040. Obviously these reductions in each of four consumption sectors will not be made in series but should occur in parallel to reach each annual five percent reduction goal.
2019 Jump to 2040	7.41 5.85	Reduce all consumption another two EBBO between 2018 and 2040 as we move toward a totally renewable-energy economy while simultaneously conserving 200 EBBO of fossil fuels.

Appendix 5

Summary of Total World Oil Production:

PAST AND FUTURE PREDICTED FOR ALL COUNTRIES RANKED BY ORIGINAL ENDOWMENT EXCLUDING OIL FROM COAL, SHALE, BITUMIN, HEAVY, DEEPWATER, POLAR, AND PLANT NGL (NATURAL GAS LIQUIDS).

All figures in Gb (Giga barrels), same as billion of barrels or EBBO as of 2002.

	Country	Past Production	Future Known and New	Total Original Endowment	Percent depleted
1	Saudi Arabia	94	206	300	31
2	Russia	121	79	200	61
3	U.S.	171	21	191	90
4	Iraq	27	108	135	20
5	Iran	54	76	130	42
6	Venezuela	46	49	95	48
7	Kuwait	31	59	90	34
8	Abu Dhabi	18	60	78	23
9	China	27	30	57	47
10	Libya	23	32	55	42
11	Nigeria	22	33	55	40
12	Mexico	30	25	55	55
13	Kazakhstan	6	34	40	15
14	Norway	16	17	33	48
15	U.K.	20	12	32	63
16	Indonesia	20	11	31	65
17	Algeria	12	16	28	43
18	Canada	19	6	25	76
19	Azerbaijan	8	15	23	35
20	N. Zone	7	9	16	44
21	Oman	7	8	15	47
22	Qatar	7	6	13	54
23	Egypt	9	4	13	69
24	India	6	6	12	50
25	Argentina	8	4	12	67
26–64	Next 38 countries combined	87	78	166	52
	World Total	896	1,004	1,900	47

Per EIA statistics, at the 2002 rate of 78 million barrels per day (28.47 EBBO annually), total depletion would occur 35 years after 2002. By 2004 this reduces to approximately 33 years. Per EIA projections, if consumption increased to 118 million barrels per day (42 EBBO) by 2025, world consumption between 2002 and 2025 would average 35 EBBO annually for 23 years for a total of 805 EBBO. By 2025 we will have used 80% of our remaining known and new reserves of 1004 EBBO.

The message is clear. In the past 150 years, we have completed the first one-half of the age of oil. At the present rate of consumption, we will compress the second half into less than 40 years. At the EIA forecasted increase rate, we will be empty in 28 years.

The previous table was reproduced by permission from page 237 of *The Essence of Oil and Gas Depletion* by C. J. Campbell, published in 2003 by Multi-Science Publishing Co., Essex, England. Tel: (01277)224632. The same numbers can be seen in visual form in the "General Depletion Picture" for oil and gas liquids published monthly by the Association for the Study of Peak Oil and Gas (ASPO).

Appendix **6**

Energy 101... Energy, Work, Power

(This Appendix segment was removed from the original Chapter 2)

The following basic definitions, energy equivalences, and necessary explanations will be helpful as background for this book or other references:

ENERGY is the ability to do work.

WORK is the movement, of something from one place to another, or the equivalent amount of heat energy required to raise something from one temperature to a higher temperature.

POWER is how fast the work happens or the rate that the energy is being used.

Work (W) can be expressed as a force (F), like a push or a pull, occurring over a distance (D) and in the direction traveled. For instance, if it takes 10 pounds of push or pull to provide enough sliding force to move a sled, and the distance traveled is 50 feet, the force times the distance equals the work.

In equation form:

Force x Distance = Work

With the numbers substituted in our sled example:

10 lbs. x 50 ft. = 500 ft.lbs.

Note, that if a 9 pound force was exerted against the sled and no movement occurred, no work would be done.

$$9 \text{ lbs. x } 0 \text{ ft. } = 0 \text{ ft.lbs.}$$

You could lean (or prop a stick in your place) against the sled all day, the sled would not move and no work would be done.

Work requires both force and distance traveled. Energy, which can be in different forms, is the ability to do the work. Power is how fast the work is done.

There are many ways that the sled could be moved, such as with a snowmobile, a dog team, cross-country skier, a winch with an electric motor, a sail, and many more, but in each case, the sliding work done and energy used is the same, 500 foot-pounds, as long as the sum of the forces resisting movement totals 10 pounds.

If there were wheels on the sled so it rolled and required only 5 pounds of force, then the work done would be 250 foot-pounds (5 lbs. x 50 ft. = 250 ft.lbs.) The wheel was a great invention to reduce the work of transportation and lessen the required energy.

Energy exists in many different forms, but in each case suggested above, the energy required, or the ability to do the work of overcoming 10 pounds of drag, is still 500 foot-pounds. It doesn't matter if the force came from a gasoline engine, an animal, a human, or the wind. Energy is that elusive something able to do the work. The units for energy and work are the same.

At this point, we will leave energy for a minute and discuss power so that we are absolutely clear about the three terms: energy, work, and power.

It doesn't take a great leap of intuition to appreciate that moving the sled in 50 seconds is a much different task than moving it the same 50 feet in five seconds. A dog or person could easily handle the 50-second task, but it would take a snowmobile or a good gust of wind with a big sail to move it the same distance in five seconds. Power is the time rate of doing the work. In equation form:

Power = Force x Distance/Time

Our original slow sled example becomes:

P = 10 lbs. x 50 ft./50 sec. = 10 ft.lbs./sec.

In the faster example such as with wind or snowmobile:

P = 10 lbs. x 50 ft./5 sec. = 100 ft.lbs./sec.

This is ten times as much power as the slower example.

The work in each case is the same, but the power is ten times greater because the time (T) to do the work changed from a task we could easily do ourselves slowly to a ten times quicker task requiring a machine stronger than most humans. Remember, the work done and the energy used while moving between two points, no matter what the time, is the same assuming the sliding friction remains the same in each case.

If we're in a hurry, we need more power. A stronger machine or an athlete can do the same work as the weaker but in less time. Power is a measure of the effort within a specific time period.

A common English unit for power is horsepower (Hp). By definition one horsepower equals 550 foot-pounds of work done in one second. To introduce horsepower to our above slower sled problem:

10 ft.lbs./sec. divided by 550 ft.lbs. per second (per Hp) = 0.018 Hp

In the faster case, the horsepower is:

100 ft.lbs./sec. divided by 550 ft.lbs. per second (per Hp) = 0.18 Hp

In real life, as speed increases, the air resistance will start to be a significant drag factor in addition to the sliding friction. Air resistance is a squared factor. Even though it might be negligible at slow speeds it increases four-fold each time the speed doubles.

In the second case, when moving the sled ten times as fast, the air resistance would be 10^2 or 100 times greater and in theory start to add to the sliding friction depending on the speed, shape, and the frontal area of the sled. Together, air resistance plus sliding friction make up a total drag force (F) and the amount of work (F x D) required to move the sled. This is why a streamlined shape becomes so important at high speed. Air resistance explains why moving things in a hurry is much more energy intensive. However, in our sled examples, the faster case of a 50-foot distance in 5 seconds is 10 ft./sec., or only about 7 mph. This is still slow enough that air drag is insignificant compared to the sliding friction so it can be ignored.

The energy needed to overcome air resistance starts to become important above 20 mph. At higher speeds, movement through air becomes the very dominate drag force resisting most of the work (energy).

In order to complete this short review of energy, work, and power, we need to clarify different forms of energy as well as to gain a quantitative understanding of the amount of work each can do.

MECHANICAL ENERGY

To review traditional mechanical terms described above:

Work = distance times force = foot pounds

Energy is equivalent to work = foot pounds

Power = work divided by time = foot pounds per second

THERMAL (THE SUBJECT OF HEAT) ENERGY

To raise a mass of material a certain temperature also requires energy. Therefore, work, the energy required, and the temperature increase of a substance are equivalent. I will try to clarify this concept with a few definitions.

The familiar term **BTU** (British Thermal Unit) is the amount of energy (or work) required to raise one pound of water one-degree Fahrenheit. To heat one gallon (about 8 pounds) of water fifty degrees would require:

8 lbs. x 50°f = 400 BTU

The BTU is a term of energy (or work) just like a foot-pound. One BTU is directly equivalent to 778 foot-pounds.

In metric terms, one **calorie** is the amount of energy (or work) required to raise one gram of water one degree centigrade. To heat the same gallon of water as above, which weighs 8 pounds (3632 grams), 50°f (27.8°c) would require:

3632 grams x 27.8°c = 100,970 calories

Obviously one calorie is not very big, so scientists sometime use the kilocalorie (1000 calories) especially as a measure of energy value of food. When this is done, a capital C may be used (not to be confused with temperature in degrees centigrade, °c).

Thermal energy may also be the output of a chemical reaction. This occurs when the complex-carbon molecules of fuel combine with oxygen to make heat (thermal energy) and simpler, lower energy, carbon molecules like carbon dioxide (CO_2) or carbon monoxide (CO). This process is called burning or combustion.

The minimum food energy required by a healthy adult in one day is about 2,000 kilocalories. Humans are very complex machines requiring large quantities of sophisticated fuels. All basic carbohydrate-fuel sources provide about 4 kilocalories/gram of energy. The average requirement of 2,000 kilocalories per day per person equals 500 grams (just over one pound) of carbohydrates per day or about 400 pounds per year. Proteins have about the same caloric-energy value and may be a substitute for carbohydrate energy. Fats have two to three times the energy value of carbohydrates.

Further reference about kilocalorie counting for food is available on the Nutrient Data Laboratory website (**www.nal.usda.gov**).

We can now see why food production, as the fuel source for our bodies, is the primary energy concern for any society. It takes considerable energy (fossil fuel, animal power, and/or human power) to produce 400 pounds of food per year per person. With today's energy intensive modern agriculture, 10 to 20 kilocalories are used to deliver one kilocalorie of food to the dinner table. Without the fossil fuels or some other alternative source of more concentrated energy, we will have to go back to subsistence lower-yield agriculture using human and animal power to provide our food.

Now that we have enough basics of scientific terms and quantification of energy, we can better appreciate the astonishing energy content of fossil fuels and reject inaccurate or misleading information. Consider the following facts straight from any technical source book, and compare the different concentrated-energy sources with the feeble output of human labor.

One gallon of oil, kerosene, diesel fuel, gasoline, fat, etc. (they're all about the same) has the concentrated combustion energy of approximately 150,000 BTU. This is a tremendous amount of stored energy and represents the ability (before efficiency losses) to do over one hundred and sixteen million foot-pounds (150,000 BTU per gallon x 778 ft.lbs. per BTU = 116,700,000 ft.lbs. per gallon) of work. We can go down to the corner gas station and buy this amount of fuel (energy) for about $1.50. Working at a typical continuous maximum of 256 BTU per hour of power (75 watts), hour after hour, a strong healthy adult would have to work 600 hours to equal this amount of energy. No wonder we live in a age of comparatively "free" energy. Even if the fuel were converted into work at a 25 percent efficiency rate (typical for fossil-fuel energy), it would still take 150 hours of steady manual labor to equal the energy in a gallon of fossil fuel.

At $1.50/gal. for gasoline, equivalent human labor to do the same work is worth about one cent per hour. Soon to be $3.00 per gallon gasoline would still make the equivalent human labor worth only 2 cents per hour.

A similar analysis reveals that a pound of coal with 10,000 BTU is equal to about 10 hours of manual labor and a pound of dry wood with 5,000 BTU per pound contains energy equal to five hours of constant manual labor.

ELECTRICAL ENERGY

Unfortunately, for the average person trying to understand the subject of energy, the electrical community has still another set of terms for energy, work, and power. The electrical engineer starts with the basic concepts:

Volts (the force, or push, like pounds in mechanical terms) times the current (in units of amperes or the quantity of electrons flowing at the speed of light) equals watts (power).

Volts x Amps = Watts

Since we have force, distance, and time (understood as speed of light), we have power. The **watt** (or kilowatts as thousands of watts) is the fundamental unit of electrical power in both English

151

and metric systems. The equivalence of electrical and mechanical power is defined as:

746 watts (0.746 kw) = one horsepower

One horsepower and three-fourths of a kilowatt are similar in magnitude. In Europe, cars are rated in kilowatts instead of horsepower. Since a kilowatt is a unit of power, it must be multiplied by time to get back to the simpler concepts of work and energy like foot-pounds. The common product of kilowatts times hours is a measure of how much we pay for electrical energy coming into our homes from the utility grid. One kilowatt (power) times one hour = one kilowatt hour:

1 kw x 1 hr. = 1 kwh

In your monthly electrical bill you will see that electrical energy (not the power) costs about 15 cents per kilowatt-hour. A typical household may use about 700 kwh per month for about $105.00. This is an incredible bargain unique to our modern, low-cost energy, industrialized civilization.

A strong human working a complete 40 hour week can only produce 75 watts of power times 40 hours, which is equal to 3000 watt hours (3 kwh) each week. This is equivalent to the thermal energy required to heat one hot shower. At electrical costs of 15 cents per kwh, a week of hard human work would be worth about 45 cents, about one cent per hour.

The above technical background is all we need to understand the remainder of this book and the magnitude of the situation we've gotten into in just the last 150 years.

The following is offered as a brief review of the history of energy:

5000 B.C.—About this time, human beings came out of the woods or savannahs as hunter/gatherers in delicate balance with nature and started to grow a little excess food energy in the form of grains. This slight surplus of energy beyond the minimum required for survival allowed the beginning of civilization. In addition, beasts of burden were domesticated to add to the amount of work that humans alone could do as well as provide concentrated food-energy as meat and milk. The agricultural revolution allowed humans to stop moving about, build villages, and multiply in population one hundred fold from about 10 million to almost a billion by the start of the Industrial Age. This early age of grain-energy reached a peak when the expansion and military might of Rome extended to the limits that grain production could support. The use of slave labor in all "civilized" countries provided incremental energy to do additional work. This practice with its detestable human rights issue is common in biblical references and continued right into the fossil-fuel era beginning 150 years ago and even to the present in some parts of the world. The U.S. was founded in vast resources and slave labor.

500 A.D.—Non-fossil-fuel machines like water wheels and windmills were built to further energy availability and give another small boost to civilization and leisure time.

1500—Civilization continued to grow slowly until wood, which was the major source of fuel for energy needs, began to be seriously depleted. Fortunately, the first use of fossil fuels in the forms of coal and peat satisfied ever increasing energy demands and kept the quality of life gradually improving. During this period the whale population was almost decimated in a few decades for the oil to be used in lighting more extravagant homes.

1775—James Watt (1736-1819) invented the steam engine and ushered in the Industrial Age by improving on the piston pump used for removing water from coal mines. This wondrous machine allowed the conversion of concentrated, previously stored energy, such as wood and coal, into work and power far in excess of what man, beast, windmill, or waterwheel could do.

1859—About the time of the Civil War when coal driven locomotives and steamships were already well established, Colonel Edwin Drake struck plentiful crude oil in Pennsylvania. This drilling technology was rapidly followed by further discoveries in the U.S.

and then eventually worldwide. Plentiful liquid crude oil eventually supplanted coal as the number one fossil-fuel energy source because of its ease of procurement, transportation, and utilization in machines which could convert its huge energy content into useful work.

1900—Beginning about a century ago, inventors found many new ways to harness the concentrated energy of free flowing oil and cleaner derivatives like kerosene and gasoline. All types of machines evolved to power industry, agriculture, and especially transportation for military, personal, and commercial use on land, sea, and in the air. There are centenarians alive today that span this entire era. It is no small coincidence that the 100-year anniversary of the airplane exactly coincides with this hydrocarbon energy epoch. Oil became an absolutely essential military force in WWI as it provided submarines, airplanes, tanks, and ships that could be refueled at sea.

The tremendous power requirements to move large masses quickly over long distances can only be provided by fossil fuels. The only exceptions are concentrated biofuels and liquid hydrogen, both requiring even larger energy inputs for their formation than they return for useful work.

1950—The use of fossil fuels including natural gas to make hydrogen culminated in more recent years as rocket fuel. This is the only way to provide the awesome power required to propel objects outward against the pull of the very strong force of earth's gravitational field. (The use of hydrogen as a concentrated fuel in lieu of fossil fuel will be discussed in more detail in other parts of this book.)

At this point I will summarize the units and equivalents used for energy, work, and power into one table sufficient for understanding the remainder of this book and other references. Wherever possible, mechanical energy, electrical energy, and work will be quantified as kilowatt hours, and power will be in kilowatts. Thermal energy will be referred to in terms of BTU's or equivalent billion barrels of oil (EBBO), with each barrel containing 42 gallons or 6,300,000 BTU's worth of energy. Keep in mind that the total thermal energy of a fossil fuel cannot be converted directly to useful electrical energy, mechanical energy, or work without an energy loss. A 25% efficiency factor for internal combustion engines and 35% for electric power plants will be used where appropriate to calculate the equivalent fossil-fuel energy. In other words, usable, secondary electrical or mechanical output is divided by the efficiency to find the required primary energy. For instance a ten-kilowatt photovoltaic system would equal a fossil-fuel, power-plant input of 10/0.35 or 28.57 kilowatts equal to 97,486 BTU/hr.

SUMMARY OF EQUIVALENTS AND UNITS:

$1,000 = 10^3$ = thousand (or kilo, K)
$1,000,000 = 10^6$ = million (or mega, M)
$1,000,000,000 = 10^9$ = billion (or giga, G)
$1,000,000,000,000 = 10^{12}$ = trillion (or tera, T)
$1,000,000,000,000,000 = 10^{15}$ = quadrillion (or peta, P)

Energy (or work equivalent)

1 kwh = 3412 BTU
1 barrel of oil = 1846 kwh = 6.3×10^6 (million) BTU
1 kilocalorie = 3.968 BTU
1 BTU = 778 foot pounds
1 foot pound = 1.356 Joules (Newton meters)
1 BTU = 0.252 kilocalories
1 kwh = 860 kilocalories
1 kwh = 2.65×10^6 (million) foot pounds
1 gallon hydrocarbon fuel = 150,000 BTU
1 gallon biodiesel = 121,000 BTU
1 pound high quality coal = 10,000 BTU

1 pound dry wood = 5,000 BTU
  (wet wood may have zero energy value)
1 cubic meter (36 cubic feet) natural gas = 36,000 BTU
1 cubic foot natural gas = 1,000 BTU
1 trillion (10^{12}) cubic meters natural gas = 36 quadrillion
  (10^{15}) BTU
1 trillion (10^{12}) cubic meters natural gas = 5.72 EBBO
  (equivalent billion barrels of oil)
1 quadrillion (10^{15}) BTU = 0.159 EBBO

Power

1 horsepower (Hp) = 550 foot pounds/second (ft.lbs./sec.)
1 kw = 1.34 Hp
1 Hp = 746 watts (0.746 kw)
1 kw = 0.95 BTU/sec.
1 Hp = 0.71 BTU/sec.
1 watt = 1 Joule/sec. = 1 Newton meter/sec.
  (metric terms used for power)

Appendix 7

A Solar-Powered Utility Vehicle (SPUV)

An alternative to fossil fuels and/or draft animals

WHY?

The problems of fossil-fuel depletion are overwhelming. Many interim solutions are proposed and confusion reigns. In this book we discussed nuclear, hydrogen, hydro, biofuels, hybrids, etc. All are on the table but none provide a smooth transition to a truly sustainable, high quality, resource and climate benign civilization. How are we going to travel? How will we feed billions of mouths without fossil-fueled agricultural or natural gas-based nitrogen fertilizer?

The only long-term answer (in addition to immediate serious conservation of the remaining fossil fuels) is to turn directly to our primary and best energy source, the sun. Unfortunately, incoming solar radiation is very dilute and requires a large area, a long time, or both to collect significant usable energy. Trees and solar panels are two familiar examples.

Photo courtesy of John Snyder www.johnsnyder.biz

What follows, and shown in the above photo is a concept which could be an answer to four basic energy needs:

- Typical agricultural functions for food production, property maintenance, and wood harvesting.

- Transport of materials and people.

- Electricity for residential use.

- Portable, self-rechargeable, electric-power source.

The working prototype has been under development for a year. This functioning vehicle provided quantitative input and encouragement that there is hope for a solar-powered, energy-sustainable future with significantly less energy.

TECHNICAL DESCRIPTION

A basic principal of the SPUV is that the photovoltaic (PV) array **travels everywhere with the vehicle**. It can be supplemented with a trailer-mounted array. This concept:

- Eliminates being stranded and/or seriously discharging the on-board battery pack.

- Eliminates charging connections and need for fossil energy.

- Allows continuous recharging even while in operation (as long as the sun shines).

Suggested optimum specifications for the SPUV are as follows. Significant deviations lose the good balance between vehicle size, battery-pack size, and array area.

- Dimensions: eight-feet long, five-feet wide, six and one-half to eight feet high over the tilted array.

- Weight: Vehicle, 1400 pounds without batteries or panels, 2,500 pounds with eighteen 12-volt DC lead-acid batteries and four panels. Trailer, 500 pounds without batteries, 1,000 pounds with batteries and panels. For agriculture purposes, weight is beneficial in the tractor.

Note: The test vehicle is a 50-year old Farmall Cub, which still has an operable internal combustion engine and a great deal of superfluous cast iron. It weighs 1800 pounds plus only one 550-pound, deep-cycle marine battery pack. An ideal new design would have an additional 550-pound lead-acid battery storage but have nearly the same loaded weight; e.g., 2,300-pound gross weight including 1,100 pounds of lead-acid, deep-cycle batteries.

- The array on the vehicle has four 175-watt Sharp panels with a total area of 130 by 60 inches or almost six square meters. The peak charging current is 5.5 amperes at 130 volts for each array or 11 amperes with the trailer. A total of 1.4 kilowatts of peak power is equivalent to almost two horsepower coming in for every hour the sun shines.

Typical deep-cycle, lead-acid batteries are seven inches by 13 inches by 11 inches high and weigh 60 pounds each. Rated Ampere Hour capacity is 105 to 125 AH with only half of this capacity available for high power use. A typical high power requirement like plowing would be 40 to 50 amperes (20 from each parallel-battery pack) at 108 volts. This equals 5.4 kilowatts or over seven horsepower for about two hours. It would take seven hours to recharge this battery pack with a two horse power array. The ideal modern design for this concept might use nickel-metal-hydride batteries for lighter weight, more performance and many more cycles of deep discharge performance.

- The motor is an advanced K91-4003 DC series wound, weighing 60 pounds, 12 inches long by 7 inches in diameter. It is rated at 10 Hp for one hour and 35 Hp peak. The drive belt is a Browning 300 H 150 tooth belt. Fortunately the PTO on the Farmall Cub is direct drive to the transmission and is driven from the rear by the electric motor through 18 to 30 tooth belt pulleys. The clutch to the I.C.E. engine is blocked open. The pulse width controller used in the test tractor is a Curtis model 1221C with a 150 amp one-hour rating. High current switching is through two Albrecht SW-180B contactors. (Contact **www.ev-america.com** or **www.kta-ev.com** for component sources).

- Speed and power: the prototype is mechanically and electrically protected at a maximum of 100 amperes (ten kilowatts or about 14 horsepower). This is considerably more than the 50 year old tractor rated as ten horsepower. All four transmission gears are used. It takes about 50 amperes to pull a seven-foot double-disk harrow or a 12 inch plow at one-fourth acre per hour as shown on the back cover of this book. These functions could be done for about two hours continuous using the double-battery pack, or one hour with a fully-charged, single nine-battery pack.

Photo courtesy of John Snyder www.johnsnyder.biz

- Stopping: ideally, the SPUV would have regenerative braking but this feature is only necessary for over-road travel. For dependable braking, a hydraulic disk brake is shown in the above close-up photo of the drive system.

Over-road travel with the trailer and third battery pack could be up to 20 mph with a modern design. Higher speeds waste too much energy through wind resistance. In bright sunlight, with regenerative braking and trailer, the ten amperes (1.2 kilowatt) solar input will provide enough power for slow speeds with little or no battery discharge.

The heavy SPUV concept is best suited for slow agricultural activities. Although the SPUV could be used for a short trip to town or to other farm sites, it is nowhere near suited to quick personal travel as the 48 volt personal-transportation vehicle (PTV). The 1,000 pound PTV described in Chapter 4 should also carry its own on-board, solar-array generator function as a portable-power source and might look like a modern golf cart for speeds up to 25 mph.

ANIMALS (HORSES, OXEN, MULES ...)

With draft animals, consider the following:

- Farmer has nearly complete control over replacement and maintenance.

- Can be used in the woods and in winter.

- Provide recyclable manure.

- Requires skills and attention for animal husbandry, shoeing, harnessing, and implements.

- About 20 to 30 thousand pounds of hay and feed are required year-round for a team plus backup and replacement animals.

SOLAR POWERED UTILITY VEHICLE (WITH ON-BOARD ARRAY)

- No food required except sunlight. Batteries must be quickly recharged, cannot be left discharged.

- Could be shared in a community residential-suburban agricultural system.

- Needs T.L.C., understanding, and periodic replacement of the lead-acid battery storage system. Lead, itself, is a finite resource similar to fossil fuels, which must be mined and refined. This is not an insurmountable problem as long as the batteries are carefully recycled, requiring little or no new lead. Each sustainable community economy should have a facility for safely recycling lead-acid batteries, of course, using solar energy. Bioplastics or even wood would eventually have to be used for the cases. Nickel-metal-hydride batteries would significantly improve the battery weight problem but at much higher cost and questionable material availability.

- Can be used to power or supplement a residential-solar system for all modern electrical needs like, lights, chainsaw, motors, refrigeration, communication, etc. Draft animals do not fit into this picture.

- Higher-speed, short-distance community travel. Up to 20 mph would require about two hours to travel to town and back with a fully-charged battery pack.

SUMMARY

In the final analysis, the SPUV might best be used as a significant compliment to a traditional draft-animal farm and minimal bio-fuels produced for farm use **only** in lieu of fossil fuels. The waste solid-residue from biofuels **must** be returned to the biofuel fields. Otherwise the soils will quickly run down in a couple of years. The SPUV **does not** deplete the soil or require fossil-fuel imput like biofuels.

From trials with the solar-powered Farmall Cub plus theoretical calculations, the SPUV concept appears doable. It might be a realistic answer to rapidly rising fossil-fuel costs and eventual complete depletion. Hopefully this exploratory work will encourage further investment and development while there is still time.

For further information visit:

http://www.mcintirepublishing.com/pages/energy.html (for Quick Time video of the test vehicle)

www.renewables.com (for electric tractor developments in California)